Five Regions of the Future

Five Regions of the Future

PREPARING YOUR BUSINESS FOR TOMORROW'S TECHNOLOGY REVOLUTION

JOEL A. BARKER, Ed.D., *hc*
and
SCOTT W. ERICKSON, Ph.D.

PORTFOLIO

PORTFOLIO

Published by the Penguin Group

Penguin Group (USA) Inc., 375 Hudson Street, New York, New York 10014, U.S.A.

Penguin Group (Canada), 10 Alcorn Avenue, Toronto, Ontario, Canada M4V 3B2
(a division of Pearson Penguin Canada Inc.)

Penguin Books Ltd, 80 Strand, London WC2R 0RL, England

Penguin Ireland, 25 St. Stephen's Green, Dublin 2, Ireland (a division of Penguin Books Ltd)

Penguin Books Australia Ltd, 250 Camberwell Road, Camberwell,
Victoria 3124, Australia (a division of Pearson Australia Group Pty Ltd)

Penguin Books India Pvt Ltd, 11 Community Centre, Panchsheel Park,
New Delhi—110 017, India

Penguin Group (NZ), Cnr Airborne and Rosedale Roads, Albany, Auckland 1310,
New Zealand (a division of Pearson New Zealand Ltd)

Penguin Books (South Africa) (Pty) Ltd, 24 Sturdee Avenue,
Rosebank, Johannesburg 2196, South Africa

Penguin Books Ltd, Registered Offices: 80 Strand, London WC2R 0RL, England

First published in 2005 by Portfolio, a member of Penguin Group (USA) Inc.

1   3   5   7   9   10   8   6   4   2

LIBRARY OF CONGRESS CATALOGING IN PUBLICATION DATA

Barker, Joel Arthur.
Five regions of the future : preparing your business for tomorrow's technology revolution /
Joel A. Barker & Scott W. Erickson.
    p.   cm.
Includes index.
ISBN 1-59184-089-9
1. Technology.   I. Erickson, Scott W.   II. Title.
T14.B362 2005
600—dc22   2005042999

Printed in the United States of America

Set in New Caledonia   Designed by Helene Berinsky

We dedicate this book to
our wives,
our children,
Kelly, Kari, Andrew, and Kyle,
and
our grandchildren,
Emily, Susannah, Sophie, Kathryn, Grace, and Matthew,
in the hope that their future is
full of promise and opportunity.

Preface

We are living in a time when the fundamental technologies that create the structure of our societies are in tremendous turmoil. A key indicator of this turbulence is the fact that when we look at the major problems of the world, we find that *we are at one of those rare times in history when we have more solutions than we have problems*. We will demonstrate this point, as incredible as it sounds, beyond a shadow of a doubt.

Coupled to this surprising claim is a second, more important assertion: that choices of whichever technologies we use to solve our problems will have a cascading effect on the way in which all of our technologies interact. Bad choices will echo painfully, not just for years, but for decades, shaping the future of businesses, communities, and countries.

This book explores the choices and describes the options that are unfolding for humanity. As futurists who have been studying this phenomenon for the past twenty-five years, we believe that we have identified five major directions the future could take. Each of these five directions is strongly driven by specific sets of technologies. We are very optimistic about these future directions and the technological clusters that accompany them. Each one affords great opportunity for new businesses and new ways of living.

Once the direction is determined, the technological solutions therein will shape a society's development for the next twenty to fifty years at a minimum. It is very likely that the long-term effects of these choices will be influential for at least one hundred years.

What seems to be equally clear is that different nations, regions, or groups can choose different directions, and that each of those directions is filled with many positive long-term benefits.

Right now, however, small groups of people are making the choices that will determine the direction of your future and our future. The irony is that many of these people don't even realize what they are doing. And the value of their choices is measured, most often, only in short-term gains. They are ignoring the long-term implications of their decisions.

This piecemeal and short-term approach is dangerous and dysfunctional. It creates the potential for mismatches of technologies that could create significant problems of efficiency and compatibility in the years to come. It also leaves out of the decision-making process the very people who are most affected by the decision.

As you read this book, you will notice that three topics are all but ignored. As we outlined this book, we realized that in order to get our ideas into one volume, we had to specifically avoid war, politics, and religion. While each of these topics is profoundly affected by technology and profoundly affects technology, each will require its own book in order to relate all of the connections. So, we will hold those commentaries for later. We believe that there is more than enough food for thought in this first book.

It is our hope that this book will stimulate strong conversation. We think it is imperative that the discussion and the choice of our future direction should be public ones.

Joel A. Barker, Ed.D., *hc*
Scott W. Erickson, Ph.D.

Acknowledgments

We want to acknowledge the help of a number of people who took time away from their work to help us with ours. We are particularly thankful to our agent, Margret McBride, for her patience with us and her encouragement; to our publisher, Adrian Zackheim, for his special vision and understanding of the project; and to our editor, Megan Casey, for her insight and suggestions for the manuscript. We want to say a special thank-you to our wives, Susan Barker and Elaine Erickson, for the years of support and encouragement that make the sustained work on this book possible.

Our perennial thanks go to the large number of individuals who write about new technology in dozens of publications that gave us the base from which we could track the innovations and changes in technology over the last twenty-five years.

Along with the intellectual debt we have to the futurists, visionaries, and technologists we acknowledge in the overview chapters of each of the Five Regions, we want to thank several of them for the friendship they shared and the personal time they spent with us over the course of our careers that encouraged our development as futurists: James Bright, Arthur Hawkins, Hazel Henderson, T. Lance Holthusen, Gary Hudson, Earl Joseph, Dennis and Donella Meadows, Graham Molitor, David Morris, Gifford and Libba Pinchot, and E. F. Schumacher.

Contents

INTRODUCTION

001 ▓
A New Look at Technology

In 1978, John Naisbitt, author of *Megatrends,* popularized the term *high tech*. By using that label, he elevated all sophisticated technology to the same status as "high priest" or "high church." It was powerful, capable of miraculous things, and to be revered. Naisbitt's label implied two categories: *high tech* and its logical opposite, *low tech*. As we move into the most technologically sophisticated century in our planet's history, it is imperative that we become more precise in our descriptions of what our technologies actually do. In a sense, we need a geography of technology so that we can better map our future. Just like locating our towns and cities on a physical map of the world, we now need to locate, on some kind of conceptual map, the blizzard of new products and processes that are appearing so we can better understand this "brave new world" of technology.

A better map with precise names and coordinates has several benefits. For instance, in the 1990s, when the U.S. investment community first tried to define the "New Economy," they were using technological labels that were far too vague—*high tech, nanotech, biotech,* and

1

infotech, for example—to be of much use. Since the long-term growth of the world's economy is profoundly shaped by the design and application of our technologies, we cannot be clear about our economic direction unless we are clear about our technological direction.

From the dawn of humanity, our tools have helped us shape the future. Now, instead of axes, bows, and fire-starting kits, we have a multitude of new things ranging from weather satellites and nuclear reactors to computer networks and CAT scanners. We have evolved from builders of simple tools to builders of complex technologies.

Today, more than ever before, our future is profoundly influenced by our technologies. Astounding as it may seem, *every day* we can count on at least one announcement about technology that is significant. Every day, scientists make discoveries that will form the base on which to develop new technologies, technologies that influence and alter every aspect of our lives.

During the past ten years, a few authors have addressed some of the key issues involving technology. Some wrote about high technology and how to market it or how to sustain it. Others wrote of specific kinds of technology, such as sustainable technology or solar technology or biotechnology. Clayton Christensen tried to explain why it was so hard to maintain leadership in technology in his first book, *The Innovator's Dilemma,* and then tried to show a successful strategy for technology leadership in his second.

What is clear in all of these writings is the significant impact technology has on modern business and society. But what is *not* clear is a way to understand the impact technology has on creating the future. Are there patterns of technology development that business, industry, government, and academia should be paying attention to? Is there some way to make sense of and give order to the deluge of new products and processes that make up our technological world?

We have been exploring those questions for almost twenty-five years by reading and cataloging technology reported in a wide variety of publications. After reading tens of thousands of articles and having

hundreds of conversations, we are ready to make a startling and revolutionary claim: We have found a new paradigm for understanding
the development of all technology. This new paradigm has a set of
simple rules that allows anyone to examine a new technology and to
map it into categories that are defined by the way that technology is
used to solve problems.

To illustrate what we mean, let us give you fifteen examples of
technology headlines selected from more than three hundred items
that we cataloged in a sixty-day period in 2004. See if you can find a
pattern in this list. For us, the pattern is obvious. By the time you
finish this book, it will be obvious to you, too.

"Atomic Powered Personal Aircraft" (*Popular Science,* June 2004)

"Life Extension through Diet" (*Fortune,* June 14, 2004)

"Cool Flames Reduce Pollution" (*New Scientist,* June 5, 2004)

"Invasive Fungi Attacking Natural Landscapes" (*New Scientist,*
June 5, 2004)

"Speedier Light Boosts Computer Chip Power" (*New Scientist,*
June 5, 2004)

"Hydrogen Helps Clean Diesel Exhaust" (*Forbes,* June 21, 2004)

"How Birds Banish Bugs from Their Nests" (*New Scientist,*
June 5, 2004)

"A Bacterium That Computes" (*Technology Review,* June 2004)

"A Website for Distributing Food to the Poor" (*Utne Reader,*
May–June 2004)

"Wearable Machines for Army" (*Popular Science,* July 2004)

"Rural Renaissance with Sustainable Technology" (*Mother
Earth News,* June 2004)

"Harvesting Energy from Waste Heat" (*New Scientist,* May 2004)

"The Power of Genetic Spam" (*New Scientist,* May 2004)

"What Happened to the Segway?" (*The Economist,* June 12, 2004)

"The Wisdom of Crowds: Book Review" (*The Economist,*
May 29, 2004)

Again, we ask, can you find a pattern in these headlines? You might answer that all of these headlines are about "high" technology. And you would be right. But, that observation doesn't tell us much.

In fact, within these fifteen examples are five very different kinds of technology. Each is driven by a set of values that has profound long-term implications for the users of these technologies. Each of these five kinds of technology aggregates into powerful patterns of use and application that will profoundly shape

the way we work,
the way we eat,
the way we move around the planet,
the way we build our homes,
the way we generate and use electricity,
the way we communicate,
the way we entertain ourselves, and
the way we take care of our health.

The goal of this book is to teach you about these five distinctly different kinds of technology and the future world each is creating. These are the *five regions of the future*.

We want to show you how to identify each of the technologies and then to understand their importance for your personal life and your professional life.

Many companies, around the world, are launching products with no understanding of the connection that they might have with one technological region or another. National policies are written that support one region over another for no good reason. And, in fact, in many cases, national policies are interfering with the forces of a free market that are trying to help these technological regions develop and compete.

Those of you working in large organizations, corporations, or government departments often find that these onslaughts of new technology blow you off-course, like ships in a gale. How do you plot a strategy, market products, provide services, write regulations, or make investments under such conditions? If all new technology is just a blizzard of novel things, how can you deal with it?

Once you understand the basic premise of this book, you will see the technological world with new eyes. To put it another way, you will be going through a paradigm shift as you read this book. New pathways will become visible. New destinations will make sense. You will have a new way to understand, appreciate, and predict where the technological world is going.

So, let us begin this journey to the five regions of the future to learn about five sets of technologies that will provide you choices between different possibilities for your future. We will be your guides, introducing you to a new way to perceive what is happening in technology development. We will take you to places that are beginning to emerge as areas of opportunity. We will give you the vocabulary and the rules for understanding how the next twenty-five years may unfold.

And with that information, you will be far better prepared to chart your own destination within the five regions of the future.

002 ▰

A New Kind of Ecosystem

> *Modern technology is like a*
> *Great Dane in a small apartment.*
> ROBERT POOL

Suppose one day you woke up, turned on the television, and heard the morning newscaster declare that a new ecosystem had been discovered somewhere on Earth. Suppose the newscaster went on to say that researchers had discovered not just one, but five new ecosystems. And these ecosystems were unlike anything that had ever appeared on the planet before.

That would be important news, because we have come to understand how crucial ecosystems are to the viability, indeed the very existence, of our world. Based on research done over the past twenty-five years, we believe that the world is witnessing the birth of *technological ecosystems constructed of human-made elements instead of biological elements.*

The first such "TechnEcology" began to emerge more than a century ago with the advent of the mass production of automobiles and steel. This TechnEcology has become so powerful that it has altered almost all of Earth's natural ecosystems. It is so dominant that no culture has been able to escape its influence. But, in the past thirty years, four other technological ecosystems have begun to emerge to compete with the first.

TechnEcologies are the inevitable result of accumulating discoveries, inventions, and innovations of human beings. These complex systems are not driven by the forces of nature, nor are they constrained by natural time. Instead, they are driven by the creative will of the human species and are capable of evolving at an amazing speed. This speed is possible because TechnEcologies operate in *tech-time*. A simple and obvious example of tech-time is the evolution of computers. Instead of taking thousands of years to evolve as natural organisms would have required to achieve the same complexity, computers moved in less than fifty years from initial emergence as a very primitive fifty-ton, water-cooled, building-sized machine to a three-pound portable device that is a million times faster and at least a million times more complex.

Because tech-time passes so much more swiftly than natural time, we could very well witness a complete technological transformation of our planet in the next hundred years.

Four of the five TechnEcologies are still very simple. But, just like any emerging biological ecosystem, with every passing day these regions, each very different from the others, grow more complex, more coherent, more self-supporting. It is fair to assume, based on complex systems theory, that at some point they will begin to self-organize and take control of their own evolution. If that happens, their power to evolve will accelerate exponentially. Because of this speed of evolution, it is crucial that human beings become conscious of what is going on now and begin to manage this emergence rather than waiting for it to happen on its own.

PICTURING A TECHNECOLOGY

It is easy to picture biological ecosystems because they are all around us. But technological ecosystems represent a new idea and are not so immediately understandable. To illustrate a TechnEcology, let us look at a small example, the automobile "ecosystem."

When the gasoline-powered automobile was invented, it was a very simple idea. But in order for it to be successful, it needed to be surrounded with a series of supporting elements—a kind of ecosystem. In no specific order, we have listed below many, but far from all, of the other elements of the technological ecosystem that developed in order for the automobile to become a successful technological species:

Easy access to fuel,
>   Which triggered the oil companies to develop better ways to
>   manufacture gasoline . . .
>   Which led to refineries scattered across the continent . . .
>   Which triggered the development of transportation modes
>   to get the gasoline to the point of need . . .

Which triggered the development of tanker trucks to carry the gasoline safely . . .

Which triggered the emergence of gas stations . . .

Which triggered the search for good corners to place those stations.

Tires for automobiles,

Which triggered the development of different kinds and sizes of tires to fill specific needs of cars . . .

Which triggered the development of distribution systems for tires and systems for patching tires so they could continue to be used . . .

Which triggered the development of tow trucks to move "injured" automobiles to where they could be fixed.

Repairing auto bodies and engines,

Which triggered the development of tools and locations for doing repairs . . .

Which triggered the development of paints and lubricants for engines (which triggered the development of filters for keeping the oil clean) . . .

Which triggered a need for a whole new set of skills and created a new job category: auto mechanic.

Road development for automobile use,

Which triggered the cement and tarmac industries to develop new ways of road building . . .

Which triggered the development of new equipment to lay the roads . . .

Which triggered the need to build new bridges . . .

Which led to new ways of building bridges out of steel and concrete.

Convenience stops for drivers and passengers of automobiles,
> Which led to roadside restaurants and motels . . .
> Which triggered drive-in eating places . . .
> Which triggered McDonald's . . .
> Which triggered the fast-food restaurant industry.

Well, you get the point. And while all that and more was happening, the automobile itself was evolving, becoming sleeker, better, faster, and safer by

Improving its lights . . .
Improving its seats . . .
Improving its ease of use with innovations such as electric starting, automatic shifting, and more powerful engines, and . . .
Improving its brakes so that it could stop better.

And the changes in automobiles and trucks and SUVs and minivans (all derivatives of the original species) continue to appear, "driving" the evolution of the automobile ecosystem.

Of course, we are being metaphorical when we say that the automobile "was evolving." Actually, humans were designing all those changes in response to market demand and engineering improvements. As demand changes, so too does the technology. And tech-time accelerates the pace of change. These days, automotive engineers use computers to design, build, crash test, and disassemble new automobile designs before a single piece of metal is bent. What used to take five years is now happening in a matter of months. These changes are people-driven, but, within the technological ecosystem metaphor, as the car "evolves," so does the system. Ford had its hundredth anniversary in 2003. Yet, even one century after the first Ford rolled off the assembly line, every time the automobile changes its capabilities, the rest of the system has to change in some way as well.

(This adaptation to change is a key aspect of ecosystems.) We need look no further for an example of demand-driven change than the SUV craze that began at the end of the twentieth century. The SUV is a result of a highly interactive, complex system, which, through innovation and adaptation, evolves to greater and greater complexity.

What is essential to understand with the automobile illustration is that it is a *very small example* of a TechnEcology. Each of the five regions of the future is already vastly more complex.

While these patterns of technological evolution look like the patterns of biological evolution, there is a fundamental difference. Biological ecosystems evolve and grow through self-organization and innovation. Their capacity to survive and adapt is *intrinsic* to a natural system. The forces for change may come from within the system or outside the system, but the responses to those forces arise from the programming *within* the living elements of the ecosystem. TechnEcologies, on the other hand, are "programmed" by human beings. All change and adaptation to internal and external forces is the design response of humans. So instead of *Deus* as designer, we have *humanus* as designer.

Now, before we proceed any further with our thesis, we need to define several of the working terms of this book.

DEFINITION OF TECHNOLOGY:

There are several traditional and accepted definitions of technology. These are:

1) *Technology as the practical application of knowledge, usually in a specific field, such as computer technology;*

2) *Technology as a means to accomplish a task using technical processes, methods, or knowledge, such as a new technology for storage of data;* and

3) *Technology as the specialized aspects of a particular area of industry, such as educational technology.*

Here is our definition, which combines much of the preceding into a single sentence:

Technology is a set of tools, techniques, and knowledge that can be used in combination or separately to solve problems.

DEFINITION OF ECOSYSTEM:

An ecosystem is a complex, integrated system made up of diverse living organisms that operate in competition and collaboration within the system's boundaries. Through biological interactions, it has the capacity to convert available resources into higher-value assets. It has the capacity for long-term survival over a variety of conditions and a tendency to grow, adapt, and evolve toward greater complexity whenever excess energy and resources are available.

All biological ecosystems are directed and constrained by their resource base. For instance, coral ecosystems receive their nutrients through a water base. A prairie ecosystem accesses its nutrients primarily through its soil. The nutrient base profoundly shapes the ecosystem that uses it.

Here are six distinguishing features of biological ecologies:

1. They have a wide diversity of living elements that are interconnected through relationships that range from predator/prey to highly mutualistic partnerships.

2. The elements within a system interact in powerful and supportive ways to maximize the advantage of available assets.

3. The organization of the system occurs, not through some kind of conscious planning, but through a series of individual interactions that ultimately improve the performance of the entire system.

4. These systems are very robust and can withstand great amounts of systemic shock and still survive and continue to grow and develop.

5. They evolve through innovation, both at the individual level and at the system level.

6. The innovations are driven through both competitive and mutualistic interactions of individual elements.

DEFINITION OF A TECHNECOLOGY:

A TechnEcology is a complex ecosystem of technology. The individual elements are made up of the tools and techniques invented by humans that interact in both mutualistic and competitive manners to increase the variety of technologies and the complexity of interaction.

Just like biological ecologies, TechnEcologies have the capacity to solve wide arrays of problems and to utilize excess energy and resources to grow into larger, more sophisticated, and more complex systems.

The nutrient base on which TechnEcologies grow is a set of values and beliefs that direct and constrain the choice of physical resources to be used in developing the technological elements within the larger system. So, the nutrient base of a TechnEcology is metaphysical, not physical.

We believe all six of the traits of biological ecosystems manifest themselves at some level in each of the TechnEcologies.

DEFINITION OF TECH-TIME:

There is one more definition and explanation required before we can begin to explore the five regions of the future. What is tech-time?

Tech-time is the time it takes to give birth to the next generation of technology.

Tech-time can run much faster than natural time, allowing complex technological evolution to occur at a much more rapid pace than that of complex biological evolution.

Even with the technological marvels of ancient and medieval times, tech-time didn't "tick" much faster than natural time because inventors were solitary for the most part and had very few tools to multiply their efforts. But following the Industrial Revolution, the advent of organized research (created by Thomas Edison), and, more recently, the wide use of computer simulation technology, with armies of researchers across the globe, it is possible to dramatically accelerate the development of any new idea. What used to take seven to ten years to develop now springs forth in less than twenty-four months. Computer processing speed has grown exponentially for more than fifty years. And with each increase in the speed of calculation, so too does tech-time accelerate.

We are now living in an era when technological evolution is only limited by the amount of intellectual effort that can be brought to bear on the specific challenge. Look at the worldwide response to the SARS epidemic of 2003. Resources were mobilized across the world to understand, identify, and find a treatment for SARS within a matter of weeks, because the global scientific community operated in tech-time.

The emergence of the five TechnEcologies and tech-time are related. In fact, tech-time is a logical evolutionary adaptation for technological ecosystems. The ramifications of these TechnEcologies and their speed of evolution are enormous and ambiguous. We believe that only by clearly identifying and understanding these new, *human-derived* ecosystems, made up of our technologies, can we begin to consciously manage their development and to decide how they should proceed.

003
A Geography of Technology

Earlier in the introduction of this book, we called for the creation of *a geography of technology* so that we can better map our future. We argued that a better map with more precise names has several benefits.

Since the world's economy is substantially shaped by the design and application of technologies, we cannot be clear about our economic direction unless we are clear about our technological direction.

The direction of the real "New Economy" cannot be clearly delineated until we understand the geography of these "New TechnEcologies."

We have identified and mapped the five TechnEcologies and given each region a name based on its dominant technological theme. Please be clear on this point: each of these regions is filled with high technology—very sophisticated and complex technologies. The five regions give us a way to parse our technology into finer and more useful categories. Here are the titles of the five regions:

The Super TechnEcology Region (ST)
The Limits TechnEcology Region (LMT)
The Local TechnEcology Region (LOT)
The Nature TechnEcology Region (NT)
The Human TechnEcology Region (HT)

CHARACTERIZING A TECHNECOLOGY

We can characterize each TechnEcology with the answers to four value questions:

What is the region's attitude toward *material wealth*?
What is the region's view of *science and technology*?
How does the region view *its relationship with nature*?
What is the region's view of *work and leisure*?

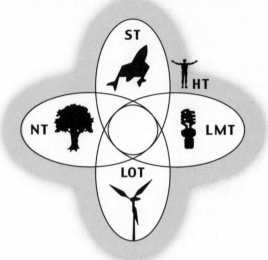

Five-Regions Venn Diagram—*We are using a Venn diagram to represent the overlapping technology between the regions*

Using these four questions, we have found that we can place almost any example of technology into a region once we know its *dominant purpose or function.* Notice that we do not say "sole" purpose or function, because most tools can be used in more than one way. We can use a screwdriver to open a bottle of beer, but that is not its dominant purpose. Once the dominant purpose of the technology is identified, we can place it in the proper technological ecosystem. This is fundamentally different from the labels we see used on technology today. Something labeled "nanotech" tells you nothing about its function. Instead, it tells you about its origin. The same is true of "infotech" or "biotech." These labels give you no understanding of their specific purpose. Ours do. This point is so fundamentally important to our thesis that it is worth restating: You cannot look at a technology and place it in one of the regions until you understand how it is going to be used. Only then can you categorize it. So, *purpose determines place.*

Many of the technologies we have cataloged are optimized for use within only one region. A simple illustration from nature is useful here: Gill technology for fish is very successful within a water environment but could never be used for its dominant purpose on land. So, it would be easy to categorize gill technology as "water" technology because of its purpose.

On the other hand, some technologies will be found in many regions. "Spine" technology is an example. The spiny urchin uses spines in a water environment, but spines can also be found on porcupines in forests and caterpillars on prairies.

If you examine the spine's purpose in each environment, you will find that the purposes are different. While the spines are used for protection in all three cases, the spiny urchin uses its spines as its dominant method of locomotion. The porcupine gains some insulation from its spines. And the caterpillar uses its spines for identification and mate selection.

So, the cataloging of the technologies in this book is based on the dominant purpose of the technology. The technological "regions" we have created have worked very well for the past twenty years in our cataloging scheme. But, we also are ready to concede that they are only a first attempt at creating useful categories. We expect disagreements with some of our cataloging. The discussions and debates that are bound to follow can only sharpen and improve our basic thesis.

One more point: Practitioners in each TechnEcology believe their region can provide, for all of humanity, a standard of living that would be considered utopian by most of the world's population. So, the choice of which TechnEcology to live in and drive your business toward is not about a better life. It is about a way of life.

The future is now beginning to unfold along lines that only science fiction writers have previously explored. We need—as a society, as a world—to examine these alternative regions, to search out the long-term positive and negative consequences, and to have public discussion, discourse, and debate about what these choices mean for our future and our children's futures. Without such discussion we run the risk of having our future determined by forces that are either too parochial or too chaotic. If we take the time, now, to explore our future, we can dramatically reduce the potential negative unintended consequences and instead shape a future full of benefits for all. What will be decided in the next ten years may well direct the next hundred years. Now is not too soon to begin the discussion.

004 ▰

Your Awareness of TechnEcologies: Taking a Measure

As we embark on this exploration of technology, we would like you to take a survey to see what your natural inclinations about technology are. There are no right or wrong answers to this survey, but our experience shows that you will find yourself responding emotionally to some of these statements. That's good! It helps us demonstrate that even though you may not have consciously been making choices about the technology in your life, you have been judging and valuing technology.

After you've completed the survey, you can then use the scoring table to discover your intuitive predisposition to the five regions. Please score the following statements using the scale below. Go to www.FiveRegionsoftheFuture.com to take the electronic version. These statements are not numbered sequentially, but don't let that bother you. It will make your task of scoring easier at the end.

1. Having cheap and abundant energy for everyone on Earth would be a good thing.

1	2	3	4	5	6	7	8	9
STRONGLY DISAGREE							STRONGLY AGREE	

12. Small, beautiful, energy-efficient homes that last hundreds of years are the best way to deal with housing needs.

1	2	3	4	5	6	7	8	9

STRONGLY DISAGREE   STRONGLY AGREE

23. Many healthcare issues can be dealt with in simple ways without the need for big hospitals and complex medicine.

1	2	3	4	5	6	7	8	9

STRONGLY DISAGREE   STRONGLY AGREE

33. There are medicines hidden in the forests and the ocean that Nature has developed that we have yet to discover.

1	2	3	4	5	6	7	8	9

STRONGLY DISAGREE   STRONGLY AGREE

46. The resources most important to human beings are traits like teamwork, leadership, and ingenuity.

1	2	3	4	5	6	7	8	9

STRONGLY DISAGREE   STRONGLY AGREE

2. Every person should have a home with more than enough room for all of his or her needs.

1	2	3	4	5	6	7	8	9

STRONGLY DISAGREE   STRONGLY AGREE

13. Spending large amounts of money on individual healthcare in the very last days of life is not worthwhile. Instead, the dying should be given care, comfort, and love.

1	2	3	4	5	6	7	8	9
STRONGLY DISAGREE							STRONGLY AGREE	

24. The number of people in the world is not the issue. It is the number of people in your neighborhood.

1	2	3	4	5	6	7	8	9
STRONGLY DISAGREE							STRONGLY AGREE	

35. Nature's many species communicate with one another in sophisticated, nonelectronic ways, most of which we don't understand.

1	2	3	4	5	6	7	8	9
STRONGLY DISAGREE							STRONGLY AGREE	

40. Using natural living systems to safely clean out infected wounds is much better than using all sorts of antibiotics.

1	2	3	4	5	6	7	8	9
STRONGLY DISAGREE							STRONGLY AGREE	

4. To make sure our species has enough space to grow and prosper, we should be willing to engineer Mars into a livable planet.

1	2	3	4	5	6	7	8	9
STRONGLY DISAGREE							STRONGLY AGREE	

15. Very sophisticated wireless telephones are the best way to bring communications to the developing world.

1	2	3	4	5	6	7	8	9
STRONGLY DISAGREE						STRONGLY AGREE		

21. It is better to have many small-scale energy systems using different energy sources than to have one large monolithic system.

1	2	3	4	5	6	7	8	9
STRONGLY DISAGREE						STRONGLY AGREE		

32. I would love to live in a home built with living systems that provide insulation, clean water, heat, and maybe even light.

1	2	3	4	5	6	7	8	9
STRONGLY DISAGREE						STRONGLY AGREE		

45. Humans communicate in ways we have yet to identify and understand.

1	2	3	4	5	6	7	8	9
STRONGLY DISAGREE						STRONGLY AGREE		

3. I believe everyone in the world deserves to have access to the most advanced healthcare technology.

1	2	3	4	5	6	7	8	9
STRONGLY DISAGREE						STRONGLY AGREE		

16. People should recycle everything they can.

1	2	3	4	5	6	7	8	9

STRONGLY DISAGREE   STRONGLY AGREE

26. I would rather buy products that are designed and manufactured locally than products that come from halfway around the world, even if they are a little cheaper.

1	2	3	4	5	6	7	8	9

STRONGLY DISAGREE   STRONGLY AGREE

49. Education is the most powerful way to give people the power to shape their own future.

1	2	3	4	5	6	7	8	9

STRONGLY DISAGREE   STRONGLY AGREE

5. Everyone on Earth should be able to access the World Wide Web anytime they want.

1	2	3	4	5	6	7	8	9

STRONGLY DISAGREE   STRONGLY AGREE

17. To quietly walk through a forest or prairie with no mechanical noises in the background makes for a great vacation.

1	2	3	4	5	6	7	8	9

STRONGLY DISAGREE   STRONGLY AGREE

22. If we had the proper skills, I would prefer to build my own house with my friends.

1	2	3	4	5	6	7	8	9

STRONGLY DISAGREE   STRONGLY AGREE

31. I would rather have us develop natural processes for producing hydrogen and methane for fuel than continue using oil pumped from the ground.

1	2	3	4	5	6	7	8	9

STRONGLY DISAGREE   STRONGLY AGREE

44. I think it would be really neat to be able to control some of my body's internal functions like blood pressure just by thinking about it.

1	2	3	4	5	6	7	8	9

STRONGLY DISAGREE   STRONGLY AGREE

7. I would love going on vacation where I can ride high-speed machines through a snow-covered forest or over ocean waves.

1	2	3	4	5	6	7	8	9

STRONGLY DISAGREE   STRONGLY AGREE

11. It is much better to become more energy-efficient and design our devices to use less energy than it is to try to find new energy sources.

1	2	3	4	5	6	7	8	9

STRONGLY DISAGREE   STRONGLY AGREE

27. Humans were born on the planet Earth and should stay here.

1	2	3	4	5	6	7	8	9
STRONGLY DISAGREE							STRONGLY AGREE	

42. Humans can make a home from many things as long as they feel safe inside.

1	2	3	4	5	6	7	8	9
STRONGLY DISAGREE							STRONGLY AGREE	

6. In order to make sure that we have all that we want, we should mine the nearby asteroids to supply additional raw materials for making things on Earth.

1	2	3	4	5	6	7	8	9
STRONGLY DISAGREE							STRONGLY AGREE	

14. Many countries would be better off if their population were smaller.

1	2	3	4	5	6	7	8	9
STRONGLY DISAGREE							STRONGLY AGREE	

25. While the Internet has some value, the best kind of communication is face-to-face.

1	2	3	4	5	6	7	8	9
STRONGLY DISAGREE							STRONGLY AGREE	

34. We humans should control our own population so that we don't crowd out the precious plants and animals.

1	2	3	4	5	6	7	8	9

STRONGLY DISAGREE   STRONGLY AGREE

43. The body has a tremendous capacity to produce its own medicines to fight sickness and disease.

1	2	3	4	5	6	7	8	9

STRONGLY DISAGREE   STRONGLY AGREE

8.   Difficult and dangerous work should be done by robots.

1	2	3	4	5	6	7	8	9

STRONGLY DISAGREE   STRONGLY AGREE

19. I would be willing to drive a very small, safe car if it got two hundred miles per gallon.

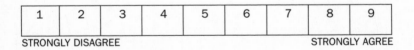

1	2	3	4	5	6	7	8	9

STRONGLY DISAGREE   STRONGLY AGREE

29. At work or at play, our obligation is to not harm the environment.

1	2	3	4	5	6	7	8	9

STRONGLY DISAGREE   STRONGLY AGREE

41. The greatest energy source in the world is human passion, determination, and genius.

1	2	3	4	5	6	7	8	9

STRONGLY DISAGREE   STRONGLY AGREE

9. Everyone having their own private jet plane for travel and vacations would be great.

1	2	3	4	5	6	7	8	9

STRONGLY DISAGREE   STRONGLY AGREE

18. Many of the most important jobs in the twenty-first century will focus on reclaiming the damaged environment.

1	2	3	4	5	6	7	8	9

STRONGLY DISAGREE   STRONGLY AGREE

30. The most important part of life is the time spent working and playing with close friends and family.

1	2	3	4	5	6	7	8	9

STRONGLY DISAGREE   STRONGLY AGREE

48. Spending time making music improves your math ability.

1	2	3	4	5	6	7	8	9

STRONGLY DISAGREE   STRONGLY AGREE

10. Our science has become so sophisticated that we can correct any environmental problems that we create.

1	2	3	4	5	6	7	8	9

STRONGLY DISAGREE   STRONGLY AGREE

20. I think that we have an obligation to stop the global warming problem by being more efficient in our lifestyles.

1	2	3	4	5	6	7	8	9

STRONGLY DISAGREE   STRONGLY AGREE

36. As soon as it is possible, we should use bacteria to mine metals from raw ore instead of sending miners underground.

1	2	3	4	5	6	7	8	9

STRONGLY DISAGREE   STRONGLY AGREE

47. Our most important work is to learn how to get along with one another without resorting to conflict.

1	2	3	4	5	6	7	8	9

STRONGLY DISAGREE   STRONGLY AGREE

50. Laughing improves general good health and increases resistance to infection.

1	2	3	4	5	6	7	8	9

STRONGLY DISAGREE   STRONGLY AGREE

28. Cars are, at best, only a partial solution to getting around. Bicycle use should be encouraged with more bike paths.

1	2	3	4	5	6	7	8	9
STRONGLY DISAGREE							STRONGLY AGREE	

37. I support making changes in animals and plants by altering DNA as long as we don't abuse them.

1	2	3	4	5	6	7	8	9
STRONGLY DISAGREE							STRONGLY AGREE	

38. I would love to be able to spend more time getting to know the animals and plants in the wilderness.

1	2	3	4	5	6	7	8	9
STRONGLY DISAGREE							STRONGLY AGREE	

39. We have been battling the natural world when we should be cooperating with it to live more gently on the planet.

1	2	3	4	5	6	7	8	9
STRONGLY DISAGREE							STRONGLY AGREE	

FIVE-REGIONS SCORING TABLE

Place the number score from the survey after each number and then add up the columns. For example, if you scored 9 for survey item 39, you would go to the Nature Tech column, find 39, and then put a 9 in the cell.

Super Tech	Limits Tech	Local Tech	Nature Tech	Human Tech
1	11	21	31	41
2	12	22	32	42
3	13	23	33	43
4	14	24	34	44
5	15	25	35	45
6	16	26	36	46
7	17	27	37	47
8	18	28	38	48
9	19	29	39	49
10	20	30	40	50
Total:	Total:	Total:	Total:	Total:

Total each column. Your highest score shows which region you currently have the most affinity for. A score over 50 in any region is a high score. Some people will have more than one high score. Multiple scores over 50 indicate a general optimism toward technology. Any score over 70 indicates a strong preference for that region. Multiple scores under 50 indicate a lack of comfort with the technology in general.

What we are trying to capture in this short assessment are your intuitive feelings toward different technologies. The goal of this book is to take you from intuition to understanding by showing you the five-regions pattern, which consists of explicit sets of technologies arranged in an entirely new way.

CHAPTERS AND SECTIONS

The remainder of the book is divided into chapters, one for each TechnEcology. Within each chapter are three sections:

1. A brief commentary on the region to identify the purpose of the technology and four fundamental rules that focus,

drive, and contain the development of technology in that region.

2. A short history of the region, a discussion of its proponents, and a set of examples that illustrate the technology. In almost all cases, the examples either are on the market right now or are in the prototype stage in the laboratories.

3. A narrative visit to the region structured as a kind of scenario to see how that region might develop over the next fifty years.

Scenarios

Scenarios have been used in business, government, and the military for the past sixty years to aid in planning and exploring the future. They tell a story or "future history" of what could happen. A famous use of scenarios occurred in the early 1970s, when Royal Dutch Shell used scenarios to forecast and then plan their response if an energy crisis arose. They were the only oil company not surprised by ensuing events.

Technology is the great destroyer of the past and creator of the future. Scenarios offer a way to talk about the possible consequences of change in technology. As you read them, remember that no one knows for sure what will happen and that each scenario is only one of a number of possibilities.

Some readers may wish to read the scenarios first to whet their appetites. Others may want to go to the examples to examine the kinds of technology in each region or to start with the overview and the rules of the region. And for those of you who want the most current examples, you can go to www.FiveRegionsoftheFuture.com. We update the examples for each region every month.

1

SUPER TECH

Any sufficiently advanced technology is indistinguishable from magic.

ARTHUR C. CLARKE'S THIRD LAW

101
Overview and Guidelines

The Super Tech region receives more media coverage than all the other regions combined, because it has the longest history and it is so dramatic and so optimistic. It is the Super Tech region that most of the world knows best. This technological ecosystem has been developing for almost one hundred years, with its greatest influence between 1944 and 1975. It was the technology of this region that first defined the ubiquitous, inescapable, and now "pop" term *high tech*.

For the first sixty-five years of the twentieth century, Super Technology was the only kind of technology that received any serious and ongoing public discussion. If you had gone to the New York World's Fair in Flushing Meadows in 1939, you would have seen a diorama of the city of the future. It was a world of giant skyscrapers,

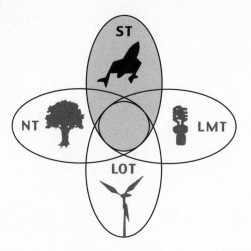

sleek airplanes, elevated roadways, streamlined cars, and enormous artificial spaces. That miniature model of a city represented one of the purest articulations of the Super TechnEcology of the time.

During the period of Super Tech dominance, we generated words to match the region:

Superman
supersonic
supermarket
superhighway
superpower
supertanker
Super Bowl
supercomputer
superstar
superhero

Supersizings, megaprojects, and big ideas were Super Tech's forte. Better, bigger, faster, and stronger were the never-ending themes.

The United States has long been the biggest member of the Super TechnEcology. We have been inside this ecosystem so long, we are mostly unconscious of its elements. But here is a short list to illustrate.

Driving an SUV is Super Tech.

If you wash your hands with antibacterial soap, you're using Super Tech.

Getting your digital TV signal from a satellite is Super Tech.

Thinking that three and a half hours' flying from Minneapolis to Los Angeles is a long time is a Super Tech attitude.

Cell phones are Super Tech. iPods are Super Tech. Fast foods are Super Tech. Having the paper towel holder in the public restroom unroll your hand towel at a touch of a button is Super Tech.

Guidelines

Here are four major tenets that frame the Super Tech ecosystem. If you are a "Super Techer," you will find the following statements reasonable.

1. Superabundance is just around the corner. Superabundance is defined as more than enough of everything for everyone on the planet. In this region no one would advocate practices like recycling or worry about resource depletion, because we'll always find more of what we need. *Around the corner* is defined as in the next twenty-five to fifty years.

2. Science and technology, given enough time and money, can solve all problems. Science and technology have driven human progress in the past. The use of science and technology is the best way to solve any new problems that appear.

3. It is time for human beings to move beyond nature with our own devices. "Thank you, Mother Nature, for getting us to this point, but we will handle it from here." Super TechnEcologists believe that they can design and manufacture anything better and quicker than Mother Nature can evolve it. We see this point of view especially in the medical arena with the advent of mechanical hearts, artificial kidneys, and electronic eyes.

4. Given the choice, human beings will always choose leisure over work. Super Techers believe humans are always seeking an easier life. We'd rather fish than farm; we'd rather watch sports on TV than exercise; we'd prefer to listen to our music on an iPod rather than teach someone to sing. In fact, all work is drudgery if you are required to do it. If we choose to do something for our own amusement, that's different. For instance, someone may choose to farm because he loves to be outdoors. But, we should use the results of our discoveries in science and technology, along with our mastery of nature, to decrease our work time and increase our leisure time. That is why robots are such important technology in this region. They will do all of the difficult, dangerous, and dreary work.

Super Techers are supremely optimistic about technology. They have seen the benefits and assume that people around the world cannot wait for the arrival of each new device.

Constant economic growth, according to the Super Techers, provides a cornucopia of material goods that raise the standard of living for people everywhere. "Ever upward and onward" is their belief. Progress is and ought to be unstoppable.

If we were to offer a single phrase to capture
the Super Tech world it might be:
"Bigger is beautiful."

102 ▩
Advocates and Examples

Advocates

The most extensive articulation of the Super Tech region can be found in science fiction from the early 1930s through the late 1970s. Olaf Stapledon, Isaac Asimov, Robert Heinlein, Arthur C. Clarke, and Larry Niven are just a few of the science fiction authors whose work is filled with the images of Super Tech.

More important, there are also a number of nonfiction authors who have emerged as creators of and spokespersons for the principles of the Super TechnEcology region.

Herman Kahn. A physicist who examined issues ranging from energy and pollution to urbanization and population, Herman Kahn often shocked people with his views of the future. For instance, if nuclear power was an answer to energy shortages and we needed the state of Montana as a nuclear waste dump to make it work, he thought that was a reasonable price to pay for progress. He stated that we would never run out of energy because we would always find new sources. After studying urban planning in Phoenix, AZ, Kahn concluded that urban sprawl and the growth of suburbs was desirable and perfectly fit the American psyche of freedom and desire for mobility. High mobility is one of the key themes of the Super TechnEcology.

In his books, *The Year 2000: A Framework for Speculation, The Next 200 Years,* and *The Coming Boom,* Kahn contended not only that ongoing American economic growth was possible and desirable but that it would continue for at least several hundred years. Kahn was not concerned about overpopulation because he was convinced that the growth curves for population worldwide would slow down before reaching the levels feared by other commentators. In the early 1970s Kahn forecast that the world population would be no higher than 6.2 billion in the year 2000. He turned out to be right and was one of the few forecasters predicting such a low number at the time. Kahn's forecasts were archetypical of the Super Tech region and were based on the belief that humans could successfully direct the growth of technology and the economy far into the future.

R. Buckminster Fuller. Best known as the creator of the geodesic dome and the man who coined the term *Spaceship Earth,* Buckminster Fuller was a prophet of superabundance. He believed that we already have more than we need of everything; we just have to be clever about how we distribute it. To solve the housing shortage shortly after World War II, Fuller wanted to build inexpensive aluminum geodesic dome homes so light that they could be transported by helicopter to wherever the homeowner wanted to live—deep forest, mountain top, or seashore. To control the negative effects of weather, he designed transparent geodesic domes large enough to cover entire cities.

Fuller was an advocate of many other "giant" projects as well, which he said would demonstrate that technology can ultimately solve all problems and that any negative result of the use of technology will eventually be replaced by another positive result of the use of technology. Like Kahn, Fuller laid the groundwork for great faith in technology and in humankind's ability to direct technology in positive directions.

Julian Simon. A professor of business administration at the University of Maryland and a senior fellow at the Cato Institute, Julian Simon is best known for his work on population, natural resources, and immigration. In his book *The Ultimate Resource 2,* Simon challenged the validity of the notions that population growth is a drain on the economy and that population growth depletes natural resources. He argued against the risk of running out of resources through over-consumption.

Simon maintained that population growth has positive economic consequences because it actually encourages economic growth. It does not deplete natural resources; it stimulates the discovery of new ones. A growing population does not result in resources becoming scarce and therefore expensive. Simon predicted in the 1970s that even with population growth, all commodities would be cheaper twenty years in the future. He was right except for oil. He also felt that a growing population had other benefits to society besides economic growth; it gives civilization more opportunities to produce more Einsteins and Beethovens.

The theories Simon developed and the data he collected helped to provide a basis for the assumptions made in the Super Tech region about abundance, growth, and long-term prosperity.

Gerard O'Neill. A physicist at Princeton University, Gerard O'Neill proposed building colonies in outer space, which is a perfect example of Super Tech thinking. The initial design for his colonies, which he proposed in the 1970s, would be spheres that rotate to produce gravity. One mile in circumference, they could house ten thousand people. The structures could be built of material found in space, either on the Moon or by capturing and mining asteroids that approach the orbit of the Earth. Once we had built the initial colonies, we could expand the concept to build cylinders large enough to house hundreds of thousands of people in utopian splendor.

O'Neill outlined his proposals in his popular book, *The High Frontier,* and gave rise to a large number of ideas about how humans could live and work in space. For example, he calculated that all the food needed to feed the entire population of the Earth could be grown in orbiting colonies that could be built from one average-size asteroid.

Since there are many thousands of such asteroids near the Earth, there is an almost limitless supply of material to meet our needs.

O'Neill pointed out that there are enough resources in outer space to satisfy our energy, food, and materials needs for thousands of years. Long term, the idea of colonizing space could not only benefit Earth directly but also provide limitless space for human settlement.

Earl Joseph. A computer scientist with Sperry Univac, Earl Joseph forecast the growth of computing speed and power accurately over the past thirty years. He was especially interested in the application of computers to devices of all sorts: smart appliances, smart homes, smart cars, and smart factories. He also saw the advent of "silicon professionals" such as doctors on a chip, managers on a chip, pharmacists on a chip, lawyers on a chip, and informational devices like a library on a chip.

Bjørn Lomborg. A very recent Super Technology advocate, Bjørn Lomborg is the author of *The Skeptical Environmentalist*, published in 2001. His work attempts to explore the long-term consequences of Super Technology. He claims that Super Tech has, on the whole, dramatically improved the world, including the natural world.

These authors and their colleagues have been some of the key voices of the Super Tech region of the future.

Examples

Now let us look at a sampling of Super Tech examples. Because we are so used to Super Tech, we forget how pervasive it is. For instance, the remote control on your television is Super Tech. Why should you waste your time and energy getting up and changing the channel, when a push of a button will do it for you? The fact that most of us have more than one hundred channels of TV to choose from is just another example of the superabundant mentality of Super Tech. If you want more examples, we have hundreds at our Web site. Even though this is just a small sampling of Super Technology, it is a safe bet that you will know about many of these ideas, because Super Tech innovations are constantly in the news.

As you look at these examples, remember that *purpose defines place in each region.* These technologies are representative of one or more of the Super Tech rules.

ENERGY

Let us recall the "superabundance" guideline. This region, as a requirement, has to produce more than enough energy for everyone. To produce that much energy, this region favors technologies with large supplies of fuel.

Fusion Power. Fusion power is the first choice to replace fossil fuels as the chief energy source. It is the Holy Grail of Super Tech energy because it uses hydrogen as fuel and produces minimum radioactive waste. The fuel of choice is deuterium (an isotope of hydrogen), which the oceans contain in vast quantities.

Calculations done at Oak Ridge National Laboratories in the 1980s to illustrate the abundance of fuel for fusion reactors indicate that if we had deuterium-based fusion power, we could give everyone in the world twice as much energy as an average American consumes today. We could then double that amount and let the world's population grow to twenty billion (about three times the current number of people), and, over the next one million years, we would consume less than 5 percent of the available deuterium in the oceans' waters. That is exactly what Super Techers mean when they talk about superabundance.

While the idea of fusion power has been criticized by most of the scientific community, the U.S. Department of Energy, reversing a fifteen-year-old position, ordered a new evaluation of fusion research in 2004.

The Tar Sands of Canada. While we are waiting for the promise of fusion to be realized, the Super Techers believe that they can still achieve more than enough energy for everyone. The Canadian Athabascan tar sands contain more than twice as much oil as the entire Saudi Arabian inventory. The oil is mixed in with sand and gravel and requires substantial processing to turn into useful petroleum. Some oil has been produced from this source for decades; in 2003, three hundred million barrels were extracted. It is an enormous project to harvest this energy, but that is no problem for this region of the future, because big projects are its forte. So, we can harvest the tar sands until we get our fusion breakthrough.

Undersea Gas Hydrates. Another energy option for the Super Tech region is undersea gas hydrates made up of methane gas compressed into solid ice under the enormous pressure of the oceans' waters. Best estimates suggest that there is enough of this fuel waiting to be harvested at the bottom of the sea to meet the entire world's energy needs for at least the next five hundred years. Recent discoveries indicate that additional hydrate deposits are frozen in the ice packs.

These three options only begin the long list of potential energy sources that Super Techers believe are available for their region of the future.

Why is superabundance of energy so important for this region? You cannot develop and deploy the rest of its technology unless you have a superabundance of energy. Without it, this technological ecosystem would collapse.

ROBOTS

Robots have become the icons of the Super TechnEcology. Robots first appeared in the stories of many popular science fiction writers such as Isaac Asimov. The 2004 movie *I, Robot* was based on stories Asimov authored half a century before. Robots can be built to do many things that people or even animals do, except that the robots can be stronger, indefatigable, and perhaps smarter so that they can undertake tasks too arduous, dangerous, or difficult for human beings. Of course, a theme in *I, Robot*, as in many previous books and movies, is that the machines (or at least some of them) ultimately run amok. But in the Super Tech region humans build, perhaps with operating rules like Asimov's Three Laws of Robotics, good machines that are our friends and helpers.

Isaac Asimov's Three Laws of Robotics are:

1. A robot may not injure a human being or, through inaction, allow a human being to come to harm.

2. A robot must obey the orders given it by human beings except where such orders would conflict with the First Law.

3. A robot must protect its own existence as long as such protection does not conflict with the First or Second Law.

Now firmly established in manufacturing (in 2002 there were close to one million industrial robots in the world), robots will soon touch all aspects of human life. Here are some examples:

Pet Robots. These have been on the market for several years. Sony brought out a robot pet dog, AIBO, that can retrieve e-mail and read it aloud. Sony's SDR-3X is a humanoid robot that can kick a soccer ball into a net. Honda's new robot called Asimo is the size of a small child and can perform simple tasks. Super Techers may not want to give up the family dog, but they can have Asimo walk the dog. And MIT grads Helen Greiner and Colin Angle of iRobot developed the My Real Baby doll robot that has sold over one hundred thousand units.

Robotic Vehicles. In the summer of 2004 robotic vehicles attempted to drive both off-road and on-road from Las Vegas to Los Angeles. The drive was a contest sponsored by the Defense Advanced Research Projects Agency (DARPA) to test autonomous vehicles. Not a single vehicle made it more than one mile. But it is the beginning. And, if you are a Super Techer, you are confident that before the decade is done, a vehicle will successfully make that trip without human help.

Robotic Skeletons. A team at Kanagawa Institute of Technology in Japan has designed and built a power suit that straps onto the wearer's back. Designed for nurses in a hospital setting, a nurse weighing 140 pounds can lift and carry a patient weighing 154 pounds. In the United States, Stephen Jacobson at the University of Utah has built a powered exoskeleton for the lower body that enables the wearer to carry massive loads without tiring. The U.S. Army is following his research very carefully as they see it as a way to augment a soldier's strength and stamina. Imagine a modern-day warrior moving in fifteen-foot strides at thirty miles an hour while carrying 150 pounds of equipment.

Microrobots. Back in Japan, electronics companies have developed a robot the size of an ant that can crawl around in power plants to inspect them and even complete repairs. The Japanese nuclear industry is expected to be among the first users when repairing leaky pipes.

Robot Helpers. The success of simple robot helpers such as iRobots's Roomba vacuum cleaner marks the beginning of a series of mobile robot appliances. These robots are not humanoid in appearance but are formed for a task like mowing the yard, guarding homes, or cleaning the carpet.

Robot Sea Ranchers. Massive sea-roaming robots are being designed at MIT and the University of New Hampshire to float underwater in major ocean currents such as the Gulf Stream. Their job will be to herd, feed, and raise large numbers of fish for harvest just as cattle ranchers currently do with steers on land.

You can see the trend just as we can. More and more robots, in more and more domains, will be designed and built to replace working human beings.

POPULATION

The Super TechnEcology has a very simple population premise: the more the merrier! Professor Julian Simon put it this way: more people increase the likelihood of more geniuses who will help us deal with the new problems of the world. So, greater population is actually an insurance policy for producing geniuses. And, because of the promise of outer space, there is always room for one more.

FOOD

When it comes to food, Super Techers want more of two things: quantity of food and variety of food. To get those two things, they focus on farming the world in ways and on a scale never done before.

Salmon Farming. Why fish for wild salmon when you can farm them? We see this Super Tech approach being used in both Europe and North America. Huge pens along the sea coast are filled with millions of trapped salmon that have been genetically modified to grow very large very quickly. Food for the salmon is automatically distributed by machines. Predators are kept from attacking the fish. Catfish get the same treatment except they're in ponds. Catfish farming can be found across the southern United States. But it is also found in Asian countries.

For a long time, chickens have been "manufactured" in factories where millions of chickens lay eggs on schedule. These chickens have been turned into "manufacturing units" with a predictable egg production schedule. Their "hen houses" are so sterile that humans entering them must put on special garb to protect the chickens from outside infection. The lighting is controlled to maximize egg production. These chickens never leave the building and almost never leave their perch. For Super Techers, this kind of manufacturing mentality represents the best way to get the most out of the food system.

Pesticides and Herbicides. Controlling pests can produce higher yields. In the Super Tech region, growers attack plant and animal pests directly with man-made pesticides and herbicides. For example, Monsanto Corporation developed a soybean that is resistant to its leading weed killer, Roundup, so that this powerful herbicide could be used to control weeds and not kill the soybeans. There is little concern about the runoff of the herbicide because, if it becomes a problem, Super Techers will simply develop another technology to fix it.

The result of this kind of food production is to make sure that not only does no one starve but everyone in the world has more than enough food—superabundance for all.

TRANSPORTATION

The Supersonic Transport. The SST was a Super Tech transportation solution of the '60s. It reflected the "faster is better" mentality of this region. Although the idea was rejected in the United States, the French and British went right ahead and built the Concorde so that wealthy customers could fly supersonically across the Atlantic. And in 2005 Airbus began building the world's largest airliner, capable of carrying 555 passengers. Both of these aircraft epitomize Super Tech.

Hypersonic Airplanes. In the twenty-first century the Super Tech focus is on developing passenger ships flying about one hundred miles above the Earth that would make any location on the planet accessible in less than two hours. NASA has been working on one they call the "hypersonic skipper." Its speed would be between 5,000–7,000 mph.

Autos. SUVs are today's Super Tech car based on their size, their energy appetite, and their capacity to go almost anywhere. They are overengineered, oversized, and overpowered. None of those adjectives is pejorative in the Super TechnEcology. More Super Tech cars are coming off the drawing boards even as you read this book. Thousand-horsepower, 180-mph Cadillacs are on display at the auto shows. Big and powerful and beautiful!

Air Taxis. *Atlantic* magazine editor James Fallows wrote a book, *Free Flight*, on the idea of easy access to airplanes and on-call flights for everyone. You can choose when and where you want to go. You always move on your own timetable and always at high speed. "An idea whose time has come," was his characterization of this kind of personalized flying. This personalization is a hallmark of the Super Tech region.

Private Orbiters. A number of entrepreneurs have entered the space race. From early pioneers like Gary Hudson to recent competitors like Brian Feeney, builder of the personal space ship, *Wildfire*, and Burt Rutan, builder of the current leading contender, *Space-ShipOne*, comes the promise of safe but cheap commercial space tourism. Rutan's *SpaceShipOne* visited the edge of the Earth's atmosphere on its maiden voyage in 2004.

Space Ships. If you look at the transportation vision of this region, you will hear Super Techers talk seriously about using antimatter fuel for space ships within twenty-five years. When matter and antimatter meet, they annihilate one another, releasing a huge amount of energy. According to research done by Gerald Smith of Synergistic Technologies, one millionth of a gram of antimatter could fuel a one-year manned mission to Jupiter. Using this almost magical fuel, you could speed between the planets at "one gravity" of constant acceleration. This means that inside the craft you would feel like you were on the surface of our planet instead of floating weightless. It also means that going from Earth to Mars would take less than seven days. "Why go to Mars?" you ask. Have patience and you will see.

Personal access, high speed, and comfort on your own terms and at your own time—this is how you move around in the Super Tech world.

COMMUNICATIONS AND COMPUTERS

Remember that the goal is to give more than enough access to everyone. So, the ultimate would be a cell phone for everyone on the planet and universal personal access to the Internet and whatever comes after. And, while we are at it, instantaneous translation into any language, which is already well along the way in the laboratories of IBM.

Fiber Optics. The laying of fiber optic cables across the surface of the planet is an early stage of this grand vision, because of the extraordinary carrying capacity of the glass fibers. One glass fiber can carry thirty-two thousand simultaneous phone messages. The use of multiple colors dramatically increases the signal capacity. Right now, there are more than ten thousand fiber bundles laid and ready to go.

Even as we go to press, the telecommunications researchers are increasing the capacity of each fiber even more. This is only one of dozens of ways to move messages around the world.

PCs. The personal computer follows the Super Tech pattern as it continues to grow more powerful. Our laptop computers now run at speeds that a supercomputer did ten years ago. In another ten years, they should be at least ten times faster and cheaper. As chip prices drop, one computer for every person on the planet is a reachable target. Add to that universal access to all the information that person needs via the next version of the World Wide Web. The end result is more than enough communicating and computing power and information for everyone—the Super Tech promise fulfilled.

HOUSING

In the United States we are already building Super Tech houses. Think about the trend in house size since 1996—ever larger, with more and more rooms designed for personal use: exercise, entertainment, computing, hobbies, gardening, and cooking. Homes as large as six thousand square feet are being purchased by empty nesters who have only themselves to shelter.

The goal in the Super Tech region would be for everyone in the world to have that much space for their personal fulfillment.

Megacities. A mainstay of the Super Tech housing theme is large structures concentrated in large cities. Japanese engineers have designed an archetypal one-building city that can hold one million people. Planned to float in Tokyo's harbor, the building is a city in itself with living units, work spaces, and recreational facilities within easy reach of all the inhabitants.

At this point, you might be thinking back to Super Tech's attitude toward population and the willingness to plan habitations for all parts of the world including currently inaccessible land and even under the oceans. You may be asking yourself, What if our growing population fills the planet?

Outer-Space Homes. Not to worry; humans will simply go off planet. Let's return to Princeton physics professor Gerard O'Neill, who worked out the problem in the 1970s and proposed building colonies in space, constructed with materials mined from asteroids and the surface of the Moon, energized with solar power or fusion power, and located at the stable gravity LaGrange Points or in lunar orbit. O'Neill envisioned thousands of colonies to be built over the next several centuries, with each settlement holding hundreds of thousands of people.

Mars. While those colonies are being built, the Super Techn-Ecologists would also be working on terra-forming Mars, changing it to become Earthlike by using large-scale engineering, in order to give it an atmosphere and make it inhabitable. This again is an example of the scale of imagination of Super Techers. Even if it would take centuries to accomplish this goal, it is reasonable for Super Techers not just to think about, but to plan and execute the molding of another planet to fit human needs.

MATERIALS

One might ask where the materials come from to build all of these amazing habitats and space ships? To any Super Techer, the answer is obvious: the asteroid belt. Once humans have cheap access to outer space, based on secure and inexpensive launch vehicles, then we have cheap access to the minerals found in the lumps orbiting between Mars and Jupiter and others in near-Earth orbit. In an asteroid just one kilometer in diameter you would find 30 million tons of nickel, 1.5 million tons of cobalt, and 7,500 tons of platinum. And, according to NASA studies, there are at least one million asteroids of that size. Other asteroids will have iron, magnesium, and titanium. The amount of metals in just a few such asteroids could provide us with all the metallic resources we need for several hundred years. And the surface of the Moon provides another inventory of materials from which we can build our Super Tech world. All that is needed is for the Henry Ford of space ships to show up and create low-cost vehicles. His name may very well be Burt Rutan, mentioned earlier, who is the designer/builder of *SpaceShipOne,* which won the $10 million prize for the first private space craft to carry a pilot and passenger into space.

HEALTH/MEDICINE

In the Super Tech region, it is "natural" to use artificial organs, because the belief is that we can outengineer Mother Nature.

Heart. The proof of this belief is clearly reflected in the most recent artificial heart implant, called the AbioCor. Invented at Abiomed, a medical technology corporation located in Danvers, MA, it has kept people alive and vigorous for several months. Patient Tom Christerson survived for 512 days with the help of the AbioCor. He had been given less than thirty days to live without the mechanical support. The AbioCor heart implant continues the ever-improving quest to provide support for weak hearts and, ultimately, replacement for failing hearts.

Eyes. At the same time, research into artificial sight is going on at the University of Wisconsin, Madison. Dr. Paul Bach-y-Rita, a physician and neuroscientist, is stimulating the tongue with electronic sensors that translate images seen by a camera to help the blind "see." The U.S. Department of Energy has been funding the Artificial Retina Project in collaboration with five major universities. Their goal is to develop a "retinal prosthesis," a kind of seeing-eye chip that would be implanted in the back of the eye and wired to the optic nerve. The intent is to have a simple but usable product on the market by 2006. "The aim is to bring a blind person to the point where he or she can read, move around objects in the house, and do basic household chores," said Kurt Wessendorf of Sandia Labs.

Like all things electronic, artificial vision will start simple and rapidly increase in power and capability.

Hearing. The focus here is on creating an artificial cochlea that could replace a damaged cochlea in a human being. Early designs were implanted as long ago as 1978. Now more than twenty thousand people have some form of electronic implant. Artificial ears, developed at the Huntington Medical Research Institute in Pasadena, CA, have been implanted in cats and hold out hope for recovering hearing for those whose auditory nerves have been destroyed.

Kidneys. Artificial kidneys are being worked on at the McGowan Center for Artificial Organ Development at the University of Pittsburgh. So far no design is small enough to fit inside the body. The user plugs into the machine and it purifies their blood. The University of Michigan has added hormone and nutrient balancing as well as blood purification to their design.

Artificial Lungs. The Hattler Respiratory Catheter, developed at the University of Pittsburgh, is an implantable "lung." About eighteen inches long, it consists of hollow fiber, porous membranes that allow for exchange of oxygen and carbon dioxide removal.

It is inserted into the vein of a leg and then passed up to the vena cava, the main vein that returns blood to the heart.

Oxygen flows through the catheter and supplies the membranes so that they can add oxygen to the blood as it flows past. While the device supplies only 50 percent of the oxygen an adult needs to survive, it can supplement weak lung function or damaged lung function well enough to let the lungs heal. It can be tolerated for only two weeks at a time.

Limbs. In Scotland, researchers have made great strides developing a bionic arm that has both hand and shoulder components. In many ways, this appendage is like a piece of a robot connected to a human. It can make complex motions triggered by electrical signals from the wearer's nerve endings. The hand is covered in soft plastic material and even has wrinkles and fingerprints.

Remember what we said about the broad use of robots in the Super Tech region? The medical industry uses them, too.

Brain-Machine Interface. Researchers at Duke University and other centers are now focusing on teaching paralyzed patients to control robotic devices with their thoughts. For patients who have lost limbs or had their spines severed, this would allow them to gain both mobility and dexterity.

But this same technology is also the beginning of an interface between a patient's brain and a computer chip in their limbs that would allow them to activate and control their muscles.

Robot Surgeons. Ultimately, these organs would all be installed by robot-assisted surgeons, using machines like the "da Vinci" surgical system, the doc-in-a-box from Intuitive Surgical of Sunnyvale, CA. At the University of Michigan, a da Vinci robot assists surgeons by providing very small "arms" and "hands" that the doctor can manipulate. The robot stabilizes the doctor's hand motions so that very precise maneuvers can be made via small incisions in the patient instead of large incisions that human hands would require. These robots are not yet autonomous; they act only under direct doctor control. But within five years, they will begin to do certain surgical procedures by themselves, if their designers have their way.

Nanodocs. Twenty years from now, if the research is successful at Stanford, MIT, and Caltech, Super Techers will develop and deploy nano-sized robots, almost too small to be seen, into our blood-stream to protect us from attack by infectious agents and do genetic corrections on our cells to keep us healthy and disease-free.

Remember, we said at the beginning of this section that the Super Tech region receives more media coverage than all the other regions combined.

The United States in particular has been the driving force for Super Tech. Japan, however, has moved furthest toward a Super TechnEcology, especially in providing leadership for megaprojects and robot development. If you look at the penetration of cell phones alone, you see Japan striding into the Super Tech region. Their ongoing commitment to nuclear energy, their pioneering of supertankers, and their bullet trains are icons of the Super TechnEcology.

Most global corporations, in terms of growth orientation and product development, are much more in line with Super Tech values than with any of the other regions' values. The reason is simple: It is the safest place to develop technology because it has such a long history. Ironically, it is the conservative nature of the global corporations that has kept them from investing in the other regions.

But that is beginning to change, and we believe that the rate of this change is going to accelerate rapidly over the next ten years. We will return to this topic in the last chapter to explain why this is true, after we finish the discussion of each of the five regions.

103 ▰

A Visit to a Super TechnEcology

Now we offer, for your examination, a scenario of a world where the Super TechnEcology is the dominant ecosystem.

Transmission over the Solar Web System (SWS)
Da: September 8, 2050

To: Buffdog3 (Grampa) @ OMC.com (O'Neill Manufacturing
     Cluster), L-5
Fr: Viking8 (Rolf) @ George School.org in Newtown, PA, USA,
     Earth

Hi Grampa!

I'm almost done with my interorbital baccalaureate extended essay. You've been a big help with the research so far and I'd like to check some additional facts with you and get your reaction to what I have written. I'm trying to create a short history of the first half of this century.

Harry Amundson smiled as he read the salutations from his grandson. He sat down by his picture window, put the e-screen on his lap, adjusted the type size, and began reading the essay Rolf had sent him. Outside, the Moon eclipsed the Sun.

Grampa, I remember you telling me how tough you had it when you were my age in 2000, having to drive a car, without robot controls, ten miles to school in bad weather. I admit that it is hard for me to believe such hardship ever existed since I have only known a world full of abundance, leisure, and opportunities.

Now that I am wrapping up my last year in high school, I want to better understand the changes that made such a world a reality for me. That's why I picked "The History of the First Fifty Years of the Twenty-First Century" as the topic for my extended essay.

Oh, one more thing, Grampa: my teachers asked me to write this report in turn-of-the-century style (I'm using a historic piece of software, Microsoft Word 2000, to experience historical communications processes as part of my paper) instead of using the Micro-Apple thought processor.

Grampa chuckled and remembered how he wrote as a young man using that same communications technology. It is so much more interesting now and faster as well to use the multimedia thought processor.

As you look down on the Earth from O'Neill One, I wonder what you like most? Mom says the Earthrises are spectacular. You see the Earth every day, so they may seem ordinary to you, but since I've only been to L-5 once, seeing such sights in person is still exciting to me.

Grampa thought about that. He knew that looking down on Earth from 240,000 miles never lost its enchantment and wonder. The Earth beyond his window was about the size of a ten-foot ball and he could see the Great Rift Valley in east Africa as the Earth rotated below. Grampa used his antique telescope to see the detail of the fault line along the valley more closely and said to himself, no, it's never ordinary.

He was in the first generation of earthlings to call outer space home. Grampa also reflected on how he almost did not have the choice of living in outer space because the space program stagnated for thirty years after the triumph of landing on the Moon. It took the shuttle Columbia *disaster of February 2003 to turn the space program around. Tragic as it was, the* Columbia *accident triggered the Bush Outer Space Vision Act passed by Congress in 2006 creating a special tax to fund a safe space program. The act taxed the wealthiest 1 percent of Americans 1 percent of their total income to invest in outer space. The act made it possible for us to return to the stars.*

THE HISTORY OF THE FIRST FIFTY YEARS
OF THE TWENTY-FIRST CENTURY
By Rolf Hendrickson

Looking back on the first half of the twenty-first century reveals quite a sight. Even though the twentieth century saw great changes in every realm of human endeavor, the twenty-first century has seen even greater changes!

The result has been the creation of a society that is reaching toward all corners of the Solar System and has produced a world of abundance and leisure with opportunities for everyone to grow to their full potential. How we got to this point is quite a story. Technology has had such a positive influence on the past fifty years that it is called the Super Tech century. But, at the turn of the millennium there were many people who questioned whether or not the future would be so rosy. Two of the main reasons for this were the energy crisis and runaway pollution. Many experts pointed out that we would run out of fossil fuels and that human behavior was adversely affecting the environment.

The CO_2 and other greenhouse gas emissions that were waste products of burning fossil fuels and industrial processes had raised the average world temperature one degree F during the last part of the twentieth century. Weather patterns worldwide were being affected, and we feared the worst. Many believed this was a temporary situation—one that ongoing technological progress would be able to address and bring under control.

Even though the decision-making process was cumbersome, in 2008 the Big Ten industrial nations committed to several simultaneous steps in energy and pollution control, which abated and then reversed the worrisome trend. The first step was to ensure ongoing supplies of fossil fuels from traditional or alternative sources. The second step was to eliminate the dangerous waste products of fossil fuels. And the final step was to launch a development program for an entirely different energy supply—one that would not cause pollution.

ENERGY AND POLLUTION

Extending oil supplies turned out to be the easiest part of the solution. The vast tar sands area of Alberta, Canada, which contained about 1.7 trillion barrels of bitumen, became the target of development by the big oil companies. They developed new, much more efficient extraction systems, which made accessible three hundred billion barrels of oil from these tar sands, more than the proven oil reserves of Saudi Arabia. This discovery had a happy consequence for Canadians; it made their nation rich. The same technology was used on the Venezuelan tar sands, but because of that nation's volatile politics, much less was done with their resources.

Even as we mined the tar sands, our engineers found a way to supplement fossil fuels with another energy source: methane, trapped in huge quantities on the ocean floor in the form of methane-ice crystals. These gas hydrates were first discovered on Blake Ridge off the coast of North America, but other deposits were quickly discovered all around the globe. Mining of undersea methane-ice began in 2009 and provided a stable fuel source for the next twenty years. Methane reserves had more energy than all the world's fossil fuel reserves combined. Of course, methane produces CO_2 when burned. But, because of its less complex carbon structure, methane pollutes far less than denser carbon fuels such as coal and oil. So, methane as a new source of energy replaced the older ones. Even using methane, however, we still needed to clean up the greenhouse gases. As usual, our technology was up to the task.

We had long known that we could sequester carbon dioxide underground or in the deep ocean. The Norwegians were the first to do it in an economical way in the late 1900s. Using oil drilling technology and, later, terra-watt lasers, we drilled far underground and piped the CO_2 safely away. Carbon dioxide pumping stations were set up to store unwanted CO_2 deep underground in old coal beds, mines, salt domes, and depleted oil or natural gas reservoirs.

We also piped CO_2 into the deep ocean. The Japanese set an example by running a pipeline down into the Marianna Trench in the central Pacific, where the greenhouse gases were pumped and then held in place several miles below the ocean's surface by the overwhelming water pressure.

Other pipelines were laid to pump the waste gases to industrial parks where CO_2 was needed for production such as the manufacture of artificial Sheetrock. For the past twenty years, humans have been stabilizing the CO_2 in the atmosphere at 1910 levels.

Grampa remembered the debates and the fears, but he was on the side of those who believed that world economic growth was needed and that energy was required to drive the growth. He could look down from O'Neill when the Earth was in darkness and see lights with the naked eye on every land mass and offshore community, the result of plentiful energy in every nation.

These carbon-sequestering programs kept the CO_2 in check while the Super Tech energy solution was coming to fruition.

By 2030, long before the methane could run out, our engineers and scientists had created a minimal-pollution energy alternative: nuclear fusion. In the momentous decision-making year of 2008, the United States and Europe committed to the Prometheus project to develop fusion power. The cost would be $1 trillion to be spent over fifteen years. As it turned out, it took twenty-five years, cost $2 trillion, and ended up involving the largest twenty-five nations of the world. But it was successful, and the first commercial fusion pilot power plant was fired up in 2030.

Now, just twenty years later, planetary energy needs are supplied almost entirely by fusion power. The fusion reactors use the hydrogen isotope deuterium. To upgrade the fusion reactors further, our engineers have developed a process to convert deuterium to tritium, another hydrogen isotope, which produces four times as much energy as deuterium.

So, hydrogen has become the primary fuel of the twenty-first century. Oil is now used strictly as a base for manufacturing plastics, lubricants, and other petrochemical products.

The fusion reactors produce electricity that is routed around the world using superconducting cables. There is so much cheap electricity that it is used to break water into hydrogen and oxygen, thus producing the nonpolluting fuel used for transportation vehicles, robots, and locations where electrical wires don't reach.

ROBOTS: FRIENDS AND WORKERS

The very prospect of plentiful energy supplies gave a huge boost to one of the major developments of the era—robots, which have become the icons of the twenty-first century. During most of the past fifty years, robots of every type have been developed to do the work that people do not want to do, especially the mundane work. We have used robots in every environment: manufacturing, housekeeping, farming, fishing, undersea and space exploration, mining, as well as any other dangerous activity. The first great personal robot maker, Susan Calvin, was responsible for taking robots out of the factory and into the mainstream. In 2015, she started with household-chore robots, but she quickly led her company, U.S. Robots, into the companion robot business, and the rest is history. Calvin's child companion robot "Bobby" became the single largest-selling machine of all time.

Bobby was introduced in 2018, followed four years later by more sophisticated 'bots. All these robots were empathetic listeners and treated children with such care that adults soon began acquiring robot companions as well. When these robots achieved what was defined as consciousness in 2025, special ethical laws, first promulgated by science fiction writer Isaac Asimov, were mandated for robot manufacturers to make sure that the robots remained friendly.

During the course of the twenty-first century, robot technology has been applied to many problems and all areas of work and leisure. All large manufacturing operations, most of which are now in outer space, are totally run by robots. Within each factory, a myriad of robotic arms and welders spin in every direction to assemble any article. These assembly plants, parked between the Earth and Moon, are huge—one hundred times bigger than the largest oil tankers built in the late twentieth century—and employ robots in every job (even repairing other robots) to make the goods we want with materials mined on the Moon and asteroids.

Very small robots monitor our civilization's physical infrastructure and make repairs in places too small or dangerous for direct human intervention.

As they assemble space stations in orbit and floating cities on the oceans of Earth, our robots keep us free from danger and drudgery. They allow us to spend most of our time doing what we want to do with our lives.

ENERGY FROM SPACE SETTLEMENTS

Of course, when they started the Prometheus project, no one could guarantee that it would be successful. So the rest of the world worked on other alternatives. Right up until 2030, when the breakthrough was made in nuclear fusion, the solar energy projects were still a viable option for the planet's energy supply.

The earliest idea for the use of space settlements was as manufacturing centers for Sunsats, orbiting energy systems that used the twenty-four–hour-per-day sunlight in outer space. Material from the Moon and nearby asteroids was used for construction of habitats and solar power stations capable of beaming energy back to Earth by microwave. The material collected from the Moon and asteroids included water, iron, nickel, aluminum, and silicon—everything needed to build space settlements, manufacturing facilities, and solar power satellites. (Of course these settlements could not exist, Grampa, without your discovery of how large numbers of people could live together in orbiting space colonies. I'm very proud of you!)

Even though we never had to use the Sunsats to beam energy back to Earth, the project was worth the cost because once they were built, they were used to provide a reliable, inexpensive, and inexhaustible source of energy beamed as microwaves or lasers to the orbiting settlements, farms, factories, and transplanetary tugs.

The plentiful energy provided by the Prometheus project had a great influence not just on robots and space settlements but on all aspects of life in the Super Tech region.

THE COMMUNICATION REVOLUTION

The cutting edge of the new economy in the twenty-first century was the revolution in communication and computers. The process of miniaturization continued in computer technology during the first half of the century. This was based on two major changes in technology. The first of these changes was made possible by using chemical processes to produce nanometer-scaled computer components. The process of growing wires and switches chemically on cheap materials like plastics was perfected by 2017. The plastic material ushered in several breakthrough products such as flexible color screens that could be rolled up for storage, wall-sized flexible plastic displays, and small digital screens so cheap that they were used for labels on merchandise.

Computer technology did not stop there. In 2019 Trans-Lucent Technologies perfected diamond chip optical computers. Not only did this development make computers much smaller and faster, but also, by combining the technology with all-optical routers invented at Dell-Bell Labs, the World Wide Web expanded to a new level of development.

The Internet, already in full bloom at the beginning of the twenty-first century, had the greatest impact on society of any communication device since the printing press. The Internet continued to evolve as technology changed until it became known as the Solar Web System (SWS; pronounced swiss). Following the rules of diversity, the one web became many webs, all with full interaction, but each with its own primary purpose and system structure. Internet traffic and e-business became the primary vehicle of commerce of the industrialized countries by 2015 and for the world as a whole by 2030, by which time there was one integrated economy, which actually stretched beyond the planet into the near reaches of outer space.

By 2015 wireless communication became available to everyone in the industrialized world who wanted it and by 2030 to the rest of the world. Everyone has a palm-sized device, which serves as a phone, Internet (SWS) connection, music player, translator, and global positioning system (GPS) navigator. Most people call it the Alota Machine (AM), because it does a lot of things. All input is by voice or scanning. Everyone has their individual digital frequency, which also is their personal security number (similar to the turn-of-the-century Social Security number; of course *social security* now means being able to live without fear for personal safety—a major issue after 9-11 of the year 2001). The AM connects people to everything they need by voice command.

What started as videophones and videoconferencing in the past century has turned into holographic projection. So we have voice, data, and holographic projection on our wrist with the AM. This form of communication utilizes the vast fiber optic networks that were laid down during the last ten years of the twentieth century. It is amazing to think that we still don't consume all the bandwidth available in these optical networks.

Replication machines are another miracle of the twenty-first century, whose beginnings started in the twentieth century. Originally the replicators were called stereolithographic machines and were used to make prototypes of solid objects out of materials that had little durability.

But all that has changed, and today a replicator can combine metal and plastics to produce final products that are as long-lived as anything built in the twentieth century. Nowadays, most small items are generated by a local replicator rather than being manufactured in one place and transported to another. This has brought new meaning to the old phrases *just in time* and *just in place.*

HOUSING: PEOPLE LIVE WHEREVER THEY WANT TO
While most people prefer to live in a city, everyone likes to get away once in a while to relax. Ninety percent of the Earth's ten billion

people live in supercities where areas of work and play are intermixed. The supercity structures are a mile high and only one mile square at the base, so their footprint is much smaller than those of cities in the twentieth century. The structures are assembled from modules manufactured in outer space out of foam metals extracted from meteors. These superstrong materials produced in outer space cannot be made on Earth because the foaming process requires the microgravity of outer space. The resulting metals are light enough to float down to the construction sites from orbit and yet they are stronger, pound for pound, than any metal that can be manufactured on Earth.

These foam metals support structures one mile square and five hundred stories high. With about two thirds of the space devoted to infrastructure, that leaves about one thousand square feet of living space for each of the five million inhabitants. Ten such structures, built around vast parks, make up a supercity of fifty million people.

There is no such thing as a traffic jam, because all vehicles are choreographed and controlled. The average workweek is twenty hours. When people have to travel any significant distance—60 miles or more—they fly.

All our dwellings are smart. Every appliance knows every other appliance. The Home Control knows our likes and dislikes and keeps track of the food we eat, the medicine we use, even our soap and cleaning supplies, and it orders what is needed when it is needed. The house cooks when we don't want to, and it waters the plants and regulates the environment. All garbage and bathroom wastes are shipped to the central plasma supertorch for incineration. The plasma torch is so hot that it breaks down everything back into simple atoms, which we can then harvest and use again.

Grampa stopped to laugh at that observation. He wasn't sure that Rolf had ever even thought of what it meant to "flush" toilets. It was all done with fresh water fifty years ago. Yes, these days are definitely better.

Even though 90 percent of the population lives in the super-cities, there are some country homes called iso-lodges. My Uncle Andrew's lodge on Ball's Bluff in Virginia is a good example. I visited him as part of my research. His lodge was constructed of foam metals, flown to the site on the bluff, and placed on the prepared foundation. The site can only be reached by hover-car. It was ready to live in one day after it was delivered. Unlike dwellings in the supercities, his lodge is outside the central power grid. Each month Uncle Andrew receives a microwave energy transmission from Earth orbit to charge the superconducting battery. It is interesting to imagine the electrons whirring around the superconducting coil forever, since the lack of resistance means the battery discharges only with use. Gone are the chemical batteries of old that ran down on their own over time.

We also have cities that float on the ocean. These settlements give ready access to the resources of the ocean and the land under the sea. Some of the floating cities are close to methane mining, fish farming, or oil drilling operations. Robots do almost all the work, but people are needed to ensure that things run smoothly. This is especially true of the entertainment and leisure industries, where many people prefer to act, sing, and have fun in person. Of the Earth's ten billion people, five hundred million live on or under the ocean.

The first space settlement, named O'Neill, was started in 2020 and finished in 2030. There are now more than one hundred with two thousand planned to be completed before the turn of the century. The first settlement was a hollow sphere that rotated to produce gravity. It was one mile in circumference and housed ten thousand people. Larger space settlements housing one hundred thousand followed five years later, and now we are building modular colonies to hold one million people each. (Again, thanks to you, Grampa.)

Humankind returned to the Moon permanently in 2010. Underground cities were built and the population has grown rapidly. There are now five million people on the Moon.

The discovery of water on Mars at the poles in 2003 led to the Mars settlement project. Planning was finished in 2019 and the first Martian city was founded in 2023. Now in 2050, Mars has four times the population of the Moon.

There are underground settlements on the large satellites of the gas giants devoted primarily to scientific study. Space settlements are also located in stable orbits near Jupiter to promote the exploitation of the riches of the asteroid belt and Jupiter's moons. A great many people have moved off planet Earth. In the year 2050 there are almost one hundred million people living somewhere in outer space.

The preliminary terra-forming studies for Venus were completed in 2048 and work should commence within fifty years. The project will take one hundred years and promises to be one of humankind's most exciting undertakings. Venus will end up looking very much like Earth.

FOOD FOR THE MANY

Our Super Technologists have focused on producing fast-growing food and creating varieties that can be grown almost anyplace to meet the tastes and desires of everyone. The land, the sea, and locations off the planet are used to grow and gather food.

In the first twenty years of the twenty-first century, the great farming regions of the world developed new drought- and pest-resistant varieties of staple crops to lessen the dependence on irrigation and pesticides.

Large-scale fish farming had replaced commercial fishing throughout the world by 2024. Since then, fish have been grown in the water like crops are grown on land. There are freshwater and saltwater fish farms. The fish are grown in controlled conditions and harvested without any chance of overfishing and diminishing the total supply of any of the varieties of fish. Sport fishing still exists on a small scale. It's pursued primarily by outdoor enthusiasts.

Manufactured foods are also popular. They are produced by cracking oil and other petrochemicals to produce the basic building blocks of food. When the process is complete, you can't tell the difference between manufactured and grown food.

Today, so much food is grown so cheaply in space that orbiting farms are the preferred providers to Earth of many food varieties. But the Earth does not want to become totally dependent on space for its food supply so for some time food production will continue on the planet as well.

Like in other areas of the economy, robotically controlled farming and fishing devices have taken the drudgery out of the work.

PEOPLE PREFER FUN

People continue to want to have time for fun and entertainment. Leisure and entertainment businesses including film, sports, television, publishing, music, hotels, 3D-O, entertainment, and theme parks together accounted for 50 percent of the global/solar GNP in the year 2047. The leisure options have become endless during the twenty-first century. These options include gaming, holography, first-run movies in real theaters instead of at home, orbiting theme parks, theme parks in the ocean, simulation rooms, holodecks, travel, and wilderness exploration.

New sports have been invented during the century, and new events and competitions have been added to the Olympics. In 2018 all existing sports were limited to "natural" athletes because body augmentation was creating an advantage too great to ignore. When an augmented athlete set the broad jump record at over fifteen meters in the 2016 Olympics, it caused a reaction. Now there are two sets of competition: one for naturaletes and one for augmentors. Each sport has a governing body to regulate the sport and define what qualifies as a natural athlete or augmentor. There is another area of sports that goes beyond augmented athletes—namely robots. It all started with a funny TV program in the 1990s called *Battlebots*.

Initially, competitions were decided by software and engineering advantages that one robot team had over the other. For a while robots played basketball, football, and other human sports. But now that robots with artificial intelligence have achieved consciousness, they are inventing their own sports. Many of these sports are quite exotic and involve activities beyond any human capacity.

Other leisure activities specialize in "I-All" experiences where one person has the experience, like walking in space or sky diving, and others experience the activity indirectly through total simulation. These I-All experiences have become very popular. There are many people who have a sense of adventure but no interest in risking life or limb to have the experience. There are other experiences, like rafting down the Colorado River, that not everyone can have because there is just not enough room on the river for more than a few people at a time. But that does not matter because the holodeck recreations based on one person or group rafting on the river are so real that anyone can feel like they have been on the Colorado.

Because the demand for these experiences continues to grow, there are more, cleverer, and increasingly interesting ways developed every year to spend our time.

EDUCATION

One of our most pleasurable activities is education. Sometimes it is hard to tell where leisure-time activities end and educational activities begin. We learn all the time, during formal instruction, arranged personal experiences, and I-All activities as part of the curriculum as well.

All human knowledge is available to us through the Solar Web System (SWS). Whenever specialized knowledge is needed, however, anyone can download an Info-Implant (everyone carries several) from the On-a-Chip series. Most popular are the module plug-ins to enhance professional and social skills such as Doctor On-a-Chip, Manager On-a-Chip, Comedian On-a-Chip, and Friend On-a-Chip. Also quite popular are academic subjects such as history, languages, math, and science. Since people change careers so often, the plug-ins make it possible to assimilate new knowledge and skills quickly.

Schooling has remained important, but the whole world is the classroom. With the knowledge available real time, schools concentrate on putting the knowledge in context so it can be used at the appropriate time.

HEALTH: A LONG AND VIGOROUS LIFE

Health and well-being were fundamentally changed by technology. Each of us has a number of nanorobots cruising around inside our bodies monitoring our well-being and making repairs and dispensing medications.

Because of these nanos, everyone has good health. This is true for two reasons. First, scientific discovery has produced new methods to treat and cure all diseases. For example, death from heart disease or cancer is almost unheard of. Second, information is available to everyone to enable people to lead healthy lives. Advancing knowledge in health science is one of the most exciting stories of this century.

Thanks to new drugs that give us strong muscles, great immune systems, and extraordinary mental health, we don't have to exercise like our ancestors did. The sciences of bio-informatics, which organized the advances in biology so that they could be more readily used, and combinatorial chemistry, which allowed up to ten thousand chemical tests to be run at one time, had a major impact on the speed and cost of drug discovery starting in 1999. There was a cure for all fatal diseases by 2039. But long before then, quality of life was being most strongly affected by the aging process itself.

We started by using replacement parts that we could make better than the originals. As the human body wears out, we now make carbon fiber knees stronger and more durable than the originals. We provide superlubrication of joints to counter the effects of aging and arthritis. We use super Kevlar fiber tendons that are stronger than the originals. Whenever needed, we can implant artificial hearts, lungs, kidneys, eyes, ears, or blood.

Repair of the human body is done as part of the annual body tune-up. On those few occasions when the nanorobots need assistance and invasive procedures are required, camera-guided, laser-wielding robotic surgeons operate and seal wounds with a sonic weld or a dab of glue.

Of course, few people need surgery because everyone wears a "doctor on a wrist." All relevant information about diseases and treatments is available on a watch-sized device that monitors blood chemistry and blood pressure and is intelligent enough to prescribe treatment. For chronic conditions requiring ongoing treatment, a microchip is implanted under the skin that releases medication electronically when needed.

TRANSPORTATION: EVERYTHING FLIES

Virtually endless supplies of cheap energy fundamentally changed transportation. The automobile industry dominated the economy during much of the twentieth century just as railroads dominated much of the economy of the nineteenth century. Air travel and space travel have had the same domination in the twenty-first century. In today's world almost everything flies. By 2020 the large airlines adopted scramjet technology and built hydrogen-burning passenger airliners that could carry four hundred people at Mach 7. They were called edge jets because after takeoff they climbed to the edge of the Earth's atmosphere on their flight path.

Of course, now in the year 2050, the latest edge jets carry one thousand people at Mach 10. Flying between any two places on Earth usually takes a maximum of two hours with more time spent waiting before takeoff and after landing than flying. Air travel, in luxury, to anywhere on the planet is available to anyone for a single, reasonable price.

Hydrogen-fuelled aerocars replaced the twentieth-century "road car" for people in all parts of the world. All are made from Ultralight, a special carbon fabric derived from nanotubes, which are extremely strong. Consequently, the passengers are the heaviest part of the car. The vehicles are powered by extremely small jet engines called nanoturbines, originally developed by the Brazilians. Like big jet engines, these burn hydrogen.

The turbines, each the size of the tip of your finger, are made cheaply by applying classic silicon chip manufacturing techniques. Five thousand are used on each car to provide safety through redundancy. So, even if dozens of engines fail, there are more than enough to keep the aerocar flying.

The aerocars are guided to any destination by using Super GPS, a system that was first instituted in the twentieth century. SGPS has navigator software that interacts with the entire matrix of weather, traffic, and safety information to determine the best route. Because of SGPS, trips are safe and relaxing. Individual flying cars give people the sense of independence and freedom that they desire. Since the aerocars fly themselves, the passengers can spend time looking out at the terrain, watching videos, or playing games.

The most exciting advances in transportation during the twenty-first century have occurred in outer space. There were two major areas of development. One development was the equivalent of the space tugboat. The space tug ships receive their power from Solarsats beaming microwave energy to them. They apply this energy to water, superheating it to produce very high thrust velocities. This technology, which came into use in 2033, allowed relatively cheap freight transport around the Solar System and provided a means of cleaning up stray asteroids that might have someday threatened Earth.

Asteroid hunting and harvesting is done with the same space tugs. The harvesting is done by pulling up to the asteroid, burrowing in until the tug reaches the center of gravity, and then pushing the "cargo" to a processing center. This combination of cargo hauling and asteroid harvesting has dramatically reduced the likelihood that Earth ever will be struck again by an asteroid "planet killer." Some say that we have fulfilled our species role by becoming guardians of the planet.

Even more startling than other developments in the past fifty years has been the development of antimatter-powered (AMP) space-ships. They have become the transportation of choice for scientists, explorers, and pioneers to get around the Solar System. Although scientists learned to make antimatter and safely store it in the early years of the twenty-first century, it took the engineers until 2038 to produce a high-volume manufacturing system for antimatter and an antimatter-powered space ship.

Because these ships can constantly accelerate at one gravity to give passengers a sense of weight, travel across the entire Solar System can be done in a matter of weeks; less than two weeks is needed from Earth to Mars and about two months to Jupiter. The exploration and settlement of the Solar System has included Lunar and Martian colonies as well as outposts in the asteroid belt and around the gas giants. It was mostly done by hardy risk takers in microwave-driven space tugs. But now, with antimatter drive, the speed of human expansion will be much faster. Everything that is important in the Solar System can be reached in less than three months. And now, in humankind's greatest adventure to date, the first starship driven by antimatter engines is being built in orbit. It will carry ten thousand pioneers and is scheduled to depart for Alpha Centari within the decade. It will be humankind's greatest adventure—greater than crossing the oceans and going to the Moon put together. The ship will leave the Solar System in 2061 and will take fifteen years to reach our nearest solar neighbor.

Life span in 2050 for all ten billion of us has increased to 180 vigorous years. That means almost all of us will still be alive when the starship returns. While we are waiting, there is always something to do in this society characterized by plenty. We have solved the problems of energy and pollution. There is more than enough of everything for everybody. The planet Earth is healthy and we are reaching for the stars. Everything is not perfect. It never will be, but we are working on it.

Well, Grampa, what do you think? You've lived through these events. Do you have any suggestions? Please call.

Love,
Rolf

2

LIMITS TECH

201
Overview and Guidelines

Living within your limits, while
boring, is also so much more civilized.

OSCAR WILDE

The second region of the future is the Limits TechnEcology. This region got its launch in 1962, when Rachel Carson published the classic, *Silent Spring*, which triggered a public debate on the long-term implications of Super Tech chemicals that were harming the environment. Her careful criticism showed that the advocates of Super Tech were not aware of or interested in the negative implications of their own technology.

But it was another book that challenged the core assumptions of Super Tech and offered a pathway toward another technological ecosystem. That book was Donella and Dennis Meadows' *Limits to Growth*, published in 1972. The first half of the book was a systematic critique of the Super Tech philosophy, its behavior, and its disturbing consequences. The authors pointed out that if our species

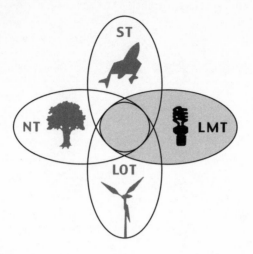

continued to do what we had been doing since the Industrial Revolution at the same rates, we would be in dangerous trouble by the end of the twentieth century.

The Super Techers' response was total rejection of the Limits Tech thesis. And the debate continues to this day. Limits Techers focus on things like overpopulation, increasing pollution, ozone depletion, destruction of fisheries and estuaries, and global warming. These, they claim, are consequences of decisions and policies that can be traced back to Super Tech behavior.

The Limits TechnEcology is predicated on a simple assumption: *The world has physical limits. When you push past these limits, sooner or later the world will push back very hard.* Limits Techers focus on creating tools and technology that permit us to have "the good life" without breaching the boundaries of the planet.

When the Limits Tech concerns were first articulated in the 1970s, they were treated with derision and ridicule by most of the world's corporations and governments.

Today, many of the "wild" ideas from *Limits to Growth* have become the norm. Citizens around the world have begun to internalize the Limits Tech ideas, whose growing acceptance has triggered the development of the Limits TechnEcology.

One of the interesting aspects of Limits Tech is the development of technology, not of things, but of practices and procedures that protect the planet from those who want to harvest everything, on both land and sea, with no consideration for the impact over the long term.

The Limits Techers have developed powerful computer models to help anticipate long-term consequences of new technologies and new behaviors. These models show that many Super Tech promises will not be "better for everyone," but, in fact, worse for most.

Here's a simple example: Antibiotics were assumed to be good in any quantity. Now we know that by using them in animals as well as overusing them in humans, we are creating superresistant bugs that are immune to the very antibiotics that were created to kill them. (Ironic, isn't it, that we call them "superbugs.") The ability and desire to test potential actions for their long-term implications before doing them is one of the key attributes of Limits Tech. They really do want to look before they leap.

Guidelines

Let us now examine the basic rules of the Limits Tech region. Here are the four guiding beliefs of the Limits TechnEcology:

1. *Scarcity is just around the corner.* All the easy-to-access resources have already been gotten. The low-hanging fruit has been picked. Soil, fisheries, forests, and petroleum, just four of many examples, have been degraded, diminished, or exhausted. From here on out we must be very careful with the resources we have left and learn to live without depleting them.

2. *Science and technology, as they have been used for the past century, are Faustian contracts, bringing short-term advantages*

but long-term disaster. Humans have been developing and applying technology without considering the long-term consequences. In a sense, we were selling our souls to the Super Tech Devil in exchange for immediate growth and benefits. But now we must develop a science of anticipation so that we can think about the long-term implications *before* we institute new technologies. Limits Techers put severe constraints on technology: It must be proven to have significant positive advantages over the long term while causing minimal or no damage to the environment.

3. *Mother Nature knows best.* Remember the saying "Don't mess with Mother Nature!" Limits Techers mean it. They assume that Mother Nature has been optimizing the environment for two billion years and has it pretty well figured out. So, when we substantially alter a natural pattern, it is likely that we are messing it up over the long term.

4. *Human beings are going to have to work very hard if they want to survive.* The planning horizon for Limits Techers is the next thousand years. They measure our actions today against how it will affect the planet ten centuries from now. And, using that measure, they believe that the human race has much work to do if we plan to survive and thrive beyond the third millennium. Much of that work will be fixing the damage we have done to Nature in the past 150 years. The rest of the work will be developing the technologies that bring us a good life while keeping Nature safe and healthy.

If there's a key descriptive phrase for the
Limits Tech region, it might be:
"Efficiency is beautiful."

202 ▮
Advocates and Examples

Advocates

Rachel Carson. As we mentioned earlier, Rachel Carson, author of *Silent Spring*, was the first popular figure to articulate the massive failures of Super Technology. In doing so, she set the challenge, which the Limits Techers picked up. Her seminal work on the impact of the pesticide DDT on birds (a Super Tech solution to malaria) alerted the world to the long-term downsides of seemingly simple technological solutions to complex problems.

DDT had wonderful short-term effect by dramatically reducing the incidence of malaria. But, over the long term, this same chemical, which had the ability to persist and accumulate in the environment, began to diminish the bird population. Rachel Carson pointed out that birds ate more insects than any insecticide could kill, plus insects could not build resistance to birds, but they could and did become immune to DDT. There were *limits* to the use of such poisons.

Paul Ehrlich. In *The Population Bomb*, Paul Ehrlich's thesis was simple: People consume resources. The more people there are consuming resources, the fewer resources per person there will be. Erhlich tried to calculate the carrying capacity of the planet and concluded that there were already too many people for optimum living, but, if we would practice zero population growth, we could stave off disaster and the human species could survive and thrive, living a simple, low-consumption life.

Garrett Hardin. In his 1968 essay "Tragedy of the Commons" Garrett Hardin described the dilemma of the environment. Simply put: Our world is based on the health of the eco-commons—the forests, the prairies, the oceans, the coral reefs, the estuaries, the atmosphere.

But, since no one group has the responsibility for taking care of these common resources, everyone tries to maximize their own benefits by overusing them. Ultimately the "tragedy of the commons" becomes a human tragedy. Almost forty years later, we still have this problem as European fishing boats trawl the African waters, taking fish that used to be harvested by Africans. Because the ocean is a "common area," they can do that legally. But the result is the over-fishing of the oceans.

Donella and Dennis Meadows. In 1972 Donella and Dennis Meadows, along with Jorgen Randers and Roger Naill, published the definitive book for this region, *The Limits to Growth.* Its essence was simple: It challenges the long-held assumption that the resources of the Earth were infinite and would always provide whatever was needed for human prosperity. Instead, the authors, through extensive computer modeling, demonstrated that humans have been growing so fast and in so many ways that in the next fifty years, we will begin violating the natural boundaries of Earth.

And thirty years later, not fifty, many of their predictions have already begun to happen. Food shortages, water shortages, resource depletion, energy cost increases—all have dramatically increased since 1972.

Using systems dynamics modeling and computers to check their assumptions, they showed how the interactions of many activities were accumulating toward collapse.

However, they never spoke of *inevitable* doom. The last sections of their book focused on the many technologies we could enlist and activities we could engage in to stop the headlong rush toward the precipice and, instead, create a healthy society with long-term success. This region is named in honor of their book.

Daniel Quinn. The author of *Ishmael* and *The Story of B,* Daniel Quinn takes up where the Meadowses left off. His two novels are examinations of the unthinking behaviors of humans as they do injury to the natural world. He also emphasizes our inability to see far enough ahead to understand the consequences of our behavior. His focus is on how to get humans back into a respectful relationship with the natural world.

Stuart L. Pimm. Pimm is the author of *The World According to Pimm*, published in 2001. His book is an audit of the natural world, and stands in stark contrast to Bjorn Lomborg's Super Tech analysis. Pimm does an especially good job at repudiating Lomborg and connecting the value of the natural world to our lives. He concludes that in many areas we are now pressing up against natural limits.

Amory Lovins. The most respected advocate for Limits Tech today is Amory Lovins. Founder and head of the Rocky Mountain Institute, he came on the scene in the late 1970s with a paradigm-shifting essay entitled "Energy Pathways: The Road Not Taken." In that essay (using data from the Electrical Power Research Institute, the research arm of the U.S. power companies), he proved that it was more profitable to conserve energy than to build power plants to generate more electricity. Since then he has helped power companies put into practice his theories on energy efficiency. He has also been a driving force in creating the "hypercar," which is designed to get very high mileage and last a very long time—both Limits Tech approaches.

Examples

ENERGY

Negawatts. Amory Lovins popularized this paradoxical and useful word. It is the measurable equivalent of a watt of energy *not* used. Lovins surveyed the wasted energy in communities. Then, through the clever use of technology, he dramatically reduced the waste. For instance, by replacing incandescent light bulbs with fluorescent lights that screw into incandescent light sockets, he reduced the energy burned per socket from one hundred watts to fifteen watts. That is a savings of eighty-five watts per socket. Although the fluorescent bulbs cost much more than incandescent bulbs, they pay for themselves in three years by their energy efficiency. And since they have a life span of five years of normal use, you end up with two more years of making money as they reduce your energy bill.

This is a key theme for Lovins: not only does the consumer save money through the efficiencies, but the power companies also do well, because they have "watts" of new energy to sell that were saved when the consumers installed their fluorescent lights. This is a win–win situation.

Do that to one million sockets and you have created eighty-five million negawatts. Continue to look for energy savings that are obvious and easy to fix, and the negawatts mount up. Pretty soon, you have saved the same amount of power that would have required an additional power plant. By not building that plant, you save the capital expense and the interest costs. Not building the plant saves all the community involvement costs. Not building the plant eliminates the pollution and other environmental damage the plant would have created. And that reduces the tragedy of the commons that Garrett Hardin wrote about. Not building the plant reduces demands on energy resources that can drive prices up.

This example epitomizes Limits Technology thinking. From the Limits Tech point of view, the advantages come not from conservation, but from efficiency. There is a huge difference between these two ideas: Conservation is usually just doing without something. Efficiency is about getting more out of the resources you have.

Through efficiency of usage, all sorts of other savings are automatically gained that have a profoundly positive ripple effect throughout an economy.

Superinsulation for Homes. In the early 1960s, the idea of heavily insulating a house was hardly even discussed. If your house was cold, turn up the heat! Now, in the twenty-first century, insulation is the norm. That is how far Limits Technology has come in the past twenty-five years. You can now buy homes that have walls and roofs constructed of panels made of a sandwich of stranded particle board and polyurethane foam (SIPS) that uses nontoxic carbon dioxide as the foaming agent. SIPS employs very sophisticated materials and manufacturing. It is definitely "high tech."

A three-thousand–square-foot house constructed of SIPS panels and proper windows and doors will consume about 80 percent less energy that an equivalent-sized house constructed in 1970 using the best technology of the day, while only costing 10 to 15 percent more. That's five times as energy efficient.

Not only do these houses save an enormous amount of energy, but they are also much more pleasant to live in, being quieter, cozier, and easier to maintain. And, because of their energy efficiency, there are other savings too. For instance, the furnace and air conditioner can be much smaller and less expensive.

While some home builders argue that the additional cost of the insulation is never fully recovered because of current low fuel costs, they do not factor in the pollution caused by the burning of the extra fuel, the increased environmental costs of mining and refining the fuel, and the potential increase of fuel costs ten to twenty years after the house is built. Limits Techers think systemically, so they are always in touch with the long-term implications.

Aerogel. Aerogel is an extraordinary insulation that weighs, in its most extreme version, just three times more than air! Some versions have an R value of sixteen per inch. The R value is a measure of a material's ability to impede the flow of heat from one side of the material to the other. An R value of sixteen per inch is quite amazing. With its additional thermal and acoustical insulation capacities, aerogel's use in refrigerators, automobiles, and homes will dramatically reduce energy use. And a new variation that is flexible is already being put into winter jackets in Italy by Grado Zero Espace.

Superconducting Cable. Superconducting materials are part of the Limits TechnEcology because electricity flows through this kind of cable without resistance. That means no loss of electrons. Present-day high-power lines lose 10 to 15 percent of their energy in line resistance. With superconducting cables, far less material is needed to carry the same amount of energy, and that energy can move around the world with almost no losses. With this technology, it might be possible to create a global energy grid and optimize the energy-producing sources, so that power plants could serve the world instead of only a region. This way, far fewer power plants, which could be environmentally optimized, would be needed to meet global energy requirements.

Microlaser Lights. These new solid-state lights—light-emitting diodes (LEDs)—were already being used as brake lights in cars in the late 1990s and in flashlights starting in 2002. They consume only 10 percent of the power of incandescent lights to produce the same amount of light and last a thousand times longer. You get a more natural spectrum of light, and you reduce your energy needs by 90 percent. Your quality of illumination is at least as good as it was before, but the energy consumption impact on the world is greatly reduced. Given the fact that electric illumination takes up about 30 percent of all the electricity generated, you can understand the "negawatts" impact of these LEDs. These lights are so durable and long-lasting that you could leave them in your will to your grandchildren.

TDP. The thermal depolymerization process (TDP) makes "waste" products into energy assets. This new technology, invented by Changing World Technologies located in Philadelphia, can take any carbon-based waste and convert it, at 85 percent efficiency, into burnable gas, fuel oil, and minerals suitable for use in manufacturing. It works with old tires, computers, knocked-down houses, and organic wastes such as pig poop and turkey guts. Before the development of TDP, all these materials were perceived as pollution that had to be properly disposed of. Now, using thermal depolymerization, all these waste products become valuable. The United States produces enough waste every year to create, through the TDP system, four billion barrels of "oil." Given that the United States consumed about 4.5 billion barrels of crude oil in 2004, the TD process fundamentally changes the oil equation. And not only does it dramatically reduce the United States' need for foreign oil, the process dramatically reduces the amount of new fossil fuel carbon released into the atmosphere, thus having a major mitigating effect on global warming.

POPULATION

Birth Control. Limits Techers are zero population growth advocates. They worry about the sheer number of people in all countries. All forms of birth control used to reduce population are Limits Tech. Scientist Paul Ehrlich has suggested that over the next several hundred years, the human population should gradually decrease its size until it reaches five hundred million people worldwide. That would be a twelvefold reduction in today's population.

TRANSPORTATION

High-Mileage Automobiles. The Toyota Prius is a current best example of a Limits Tech automobile. Designed to deliver over fifty miles per gallon while comfortably carrying four people, it sets the baseline of what is possible in high-mileage, low-emission, hybrid technology. Limits Techers are not so zealous as to believe that people will give up automobiles. So, instead they focus on making cars more efficient. Right now, in automobile research facilities around the world, there are designs that can carry four people in comfort while getting one hundred miles per gallon. That efficiency would extend our oil supplies two to five times longer and would dramatically reduce pollution in every major city in the world. The ultimate goal with cars would be one that is safe and capable of carrying four people, while getting two hundred miles per gallon.

Trains. There are two examples of Limits Technology applied to trains. The first is based on readily available technology. Argonne National Laboratory in Illinois has developed a system of "laser glazing" for the sides of railroad rails. Smoothing out this part of the rail reduces the friction between the side of the rail and the wheel by as much as 40 percent. This is a significant energy savings for the train and reduces stress fracturing of the rail itself by as much as 75 percent. This produces a cost saving and time saving in replacement and materials by extending the life of each rail.

The second train example is much more sophisticated. Remember that Limits Techers want to more efficiently use assets we already have in place. So, how can we increase the speed and efficiency of present-day trains? Sandia National Labs in Albuquerque, NM, has designed a high-speed train with a gas turbine engine that spins an electric generator to produce electromagnetic pulses, which push against an aluminum reaction rail to move the train forward. You keep the standard track and just add the aluminum "third rail." With this system a train could travel up to two hundred miles per hour on existing track. Of course, with proper streamlining to reduce air resistance, you end up with a much more energy-efficient passenger rail system at minimum additional costs. Here again is the Limits Tech approach: save what you have; improve the design to make it more efficient for the whole system, not just one part of the system.

Electronic Travel. For most travel, Limits Techers would rather have you stay at home and communicate instead of travel. Substituting electrons or photons for actual movement is the most efficient way to "travel" in the Limits Tech world.

HOUSING

Homes in the Limits TechnEcology would be much smaller than we presently see in countries like the United States. They would be built higher, not wider. They would be built closer together. They would be very energy efficient and be designed to last a long time— at least two hundred years, with a design that makes them easy to fix and upgrade. That, by the way, has been both an Asian and a European approach for the past three hundred years.

MATERIALS

Substitution. Use less, use less, and then use even less. In *The Limits to Growth,* the Meadowses talk about substituting computer storage for paper files. In 2001 an IBM microdrive, which is just twice the size of your fingertip, holds one gigabyte of data. Apple's 2004 version of its iPod music player has a one-inch disk that holds twenty gigabytes of music. Soon IBM, or some other vendor, will begin to deliver terabyte disks with one thousand billion bytes. That will hold tens of millions of pages of information on a platter of exotic material that weighs, at the most, three ounces. The Limits people support this solution because it solves the information storage problem efficiently by using "high" technology instead of clear cutting entire forests for paper.

Recycling. Limits Techers reduce their use of raw material by recycling. BMW is applying the Limits Tech philosophy by designing their cars to make them easy to recycle. Germany and other European countries have "return" rules for major products like cars and major appliances that require the manufacturers to be responsible for the full cycle from manufacturing to "internment" of products. That's Limits thinking.

Carbon Fiber. This material is very strong and very light and is made from the most common molecule on Earth. Limits Techers love it for all three reasons. Because of its strength, you don't need to use a lot of it; because it is light, it puts less stress on the other components of whatever product it is being used in; because it is common, we can manufacture it without depleting key resources. Carbon fiber began its product life in very expensive bicycles to keep them light. Then it moved into tennis rackets for its strength and lightness. Boeing's new 7E7 (the *E* stands for *efficiency,* by the way) will have many wing panels and body panels with carbon fiber in them. As more and more carbon fiber is used, it will drive the price down even further until we see automobiles made of the material.

Co-Products. Using waste from one industry to produce useful products in a different industry is another way to recycle. Carbon dioxide plus coal ash under high pressure create the equivalent of Sheetrock, according to research done by the Tennessee Valley Authority using waste from its coal-fired power plants. This process takes two major waste products and turns them into a valuable product. We end up sequestering carbon dioxide so that it won't add to global warming, we remove the ash (which otherwise has to be buried somewhere), and we create a superior building material that doesn't use up virgin materials. That's Limits Tech thinking.

COMMUNICATIONS
The Limits Tech attitude toward communications looks, at first glance, identical to Super Tech's. But, in the Limits TechnEcology, communications is used as a substitute for transportation as its dominant purpose rather than as a source of superabundant information.

Fiber Optics and Copper Wire. Instead of laying down lots of fiber optic cable to every household, Limits Techers would first work to maximize the copper wire we already have in place. For instance, they might apply the new "chirp" technology, which increases the number of signals you can place on any one frequency by a factor of one hundred. This would mean that far more complex signals could travel through on standard wires, thus avoiding any expense of new installation or change in the materials needed to supply homes with efficient communications.

The fiber optics that has already been put in place around the world would be utilized by Limits Techers in every way they could think of. Because so much fiber is already in place, it is highly unlikely that, in a Limits TechnEcology, there would ever be a need to add more fiber, except to locations that still don't have it to start with.

Books. What does a Limits Tech book look like? It looks like an "e-book," an electronic book with a flat screen and cheap electronic memory large enough to hold two hundred to four hundred books. These innovations feel like Super Tech until you examine their dominant use.

E-books save trees by eliminating the paper and cardboard boxes necessary to print and package the books. No physical books means no warehouses needed to store the books, not to mention the trucks and the fuel they require to transport the books. E-books allow a reader to download books into one device, which can store thousands of pages of text as well as newspapers and magazines—all available in one device.

If you add up the amount of material it takes to create one e-book and then subtract the amount of material it would take to match its capacity and flexibility in paper products, you find that the e-book is a wonderful actualization of the Limits philosophy. All that is needed, technologically, to achieve this e-book is an inexpensive, high-definition screen. And a recent Sony product released in Japan in late 2004 meets this requirement.

Satellites. Communications satellites are the way to go for developing countries because you don't have to put up the wire, which, in turn, saves copper and telephone posts and all the time, energy, equipment, and money needed to string that wire. And it also eliminates the maintenance costs that go along with a land line system— as well as the time it takes to replace storm-damaged lines.

Satellites with sensors that monitor the world's environment are just as important as communications satellites for Limits Techers. They wish to have very high levels of information about the conditions of the planet. This information is necessary to monitor the overall health of the planet and make sure that natural limits are not being violated.

Little Satellites. Of course, Limits Techers will look for the most efficient way to deliver communication and sensor services. For instance, "pico" satellites are very small satellites weighing less than one kilogram. The idea is to send up a swarm of them into orbit. They would rendezvous in space and operate in concert to do complex tasks.

Were you wondering about all the polluting rockets necessary to send those satellites into orbit? The solution in the Limits Techn-Ecology is to shoot them into orbit with very large electromagnetic cannons that accelerate the satellites with magnetic force instead of chemical fuels. This kind of launching process will be much cheaper, use far fewer resources, and protect the atmosphere from pollutants that chemical rockets create in their fiery exhaust.

FOOD

Preservation. The contrast between Super Tech food and Limits Tech Food is profound. Where Super Tech is always focusing on growing more food and more man-made variations of it, Limits Techers think in the opposite direction. Their most important single question is "How can we better preserve what we already harvest?"

UN data shows that almost every year the world grows more food than it needs. But during harvest and storage, we lose large amounts to pests and rot. If we could stop that, we would have surpluses in most parts of the world.

So, one Limits Technology strategy is irradiation of food. This can be done with gamma rays or electron beams. Either treatment stops spoilage in its tracks and allows properly containerized foods to remain fresh for years. If used properly, it allows fresh fruits and vegetables to last much longer as well. Even today it is being used on herbs and spices to maintain their freshness.

Farming. The growing of food too is influenced by Limits Technology. John Deere has achieved a careful mapping of farming soils coupled with precise location of the nature of those soils through global positioning satellites (GPSs). This allows the farmer to custom-fertilize his soil based on precise directions from his tractor's GPS system. The end result is a much lower use of fertilizers, insecticides, and herbicides. At the end of the year, that same GPS system maps out the levels of productivity of the farmland to fine-tune the process for the following year. Again, what we see is an attempt to limit the waste of resources, protect Nature against the overuse of chemicals, and increase productivity through efficiencies.

Ultimately, a Limits Tech farm would be completely organic, using no artificial fertilizers or insecticides or herbicides, and producing its own fuels from the use of waste materials.

Fishing. Fishing has been a primary source of food for many countries in the Third World. One approach to fishing is vehemently resisted by Limits Technologists. Called drift net fishing, it is one of the worst examples of Super Technology. These nets, which may be tens of miles long, trap not just food fish but all sorts of marine life including dolphins, turtles, and even small whales. Recent data suggests that the life in the oceans is being threatened in a fundamental way by this superfishing. All fishing may have to stop for some period of time in order to let the oceans regain their vigor.

One Limits Tech solution to this vast destruction by these nets—proposed by Donella Meadows, co-author of *The Limits to Growth*—is to outlaw the nets and go back to hand lines and rod and reel, catching tuna and other fish the way they used to be caught. Her calculations showed that such a change would create thousands of new jobs, make smaller ships more efficient than larger ships, and restore the health of the oceans all at the same time.

HEALTH

Prevention. Limits Techers advocate prevention and wellness. It's a lot cheaper to stop illness before it starts—Public health measures are the best investments. Limits Techers are big on vaccines. That makes sense—Why get sick in the first place?

Limits Techers advocate minimum expense at the end of life. Right now 20 percent of healthcare costs in the United States occur in the last ten days of life. A Super Techer would say that's fine. A Limits Techer would say there has to be a more thoughtful, less expensive way to let life end.

Nutrition. Eating properly would be considered smart medicine in the Limits TechnEcology. This goes back to the "Mother Nature knows better" belief. If we eat a wide variety of foods, cutting back especially on animal protein, we better utilize the resources available to us. The latest research on healthy foods reinforces the Limits Tech diet.

Anti-antibiotics. Limits Techers avoid antibiotics whenever possible. Why? Again, it is the "Don't mess with Mother Nature" reason. When we use antibiotics needlessly, we train bacteria to develop better ways to resist those drugs and thus make them deadlier. Limits

Techers would rather have you develop a healthier immune system. Only when the need was clearly there would antibiotics be offered.

COMPUTER SIMULATIONS

Limits Techers are big on computer simulations. For instance, right now the auto companies worldwide can model and test an automobile on a computer in such a way as to maximize its strength while reducing the materials needed to build it. We can also model its aerodynamic qualities, thus designing a more energy-efficient vehicle. We can model the engine and the combustion patterns inside the piston cylinder to make it burn its fuel with less pollution. And all of this is done without ever having to construct the real thing. Limits Techers want to study electronic prototypes of products before building them to explore the long-term implications of a new design.

In the same way, they want to do economic projections to understand the impact of new products, new policies, and new ideas on all levels of society. They want to study potential ecological decisions with computer models to make sure that humans disturb the natural world as little as possible while we are thoughtfully taking care of our needs. For all of these reasons and more, Limits Techers love simulations.

In all their technologies, Limits Techers strive for a good life without damaging the natural world.

203
A Visit to a Limits TechnEcology

Here again is a visit to one possible world fifty years into the twenty-first century. From the many technological options, we have picked this set to illustrate how a Limits TechnEcology might appear. These comments—from the journal notes of a fifty-year-old educational consultant—are written to be read by her grandchildren in thirty years.

Tomorrow morning will be the launch of the 2050 Fashion Parade. My teenage son, Mike, and my daughter, Molly, are so excited. This marks the five-year interval for fashion design. For the next month all the great design houses of the world will be rolling out their 5-Fashions. For the next thirty nights there will be 3D specials to watch.

Every year, the world has a design parade. Last year, it was for automobiles. Next year, it will be for small appliances. It is part of the Rational Limits cycle that started in 2010 to help control the high-speed obsolescence craze that climaxed in the first years of the twenty-first century.

The world decided to slow down after we realized how much wasted effort was being expended in trivial improvements of products and services that were already very good. So the Design Parade was proposed. No one was required to go along, but it just gathered momentum. When the rock group U2 and some star athletes began to support it, that made a huge difference. It became the thing to do.

As it became "fashionable" to be part of the Parade, the economics began to shift. A "virtuous spiral" began to build, and within thirty-six months, it became more economical to be part of a parade than to try to bust the pattern.

The Parade concept is very simple in principle: Each major category where technology and design are important has a five-year window. You bring out your design and then you stand by it for the next five years. At the next Parade, you can introduce your improved product, and then that has a five-year life cycle.

The only way you can break the cycle is if you significantly improve your product or service as measured in one of two ways: your new product has to be either 100 percent better in its utility than what it replaces or 50 percent cheaper in all cost measures. Either of those two improvements lets you into the marketplace immediately. Needless to say, that doesn't happen very often.

So, tomorrow we will begin to see what the new look in clothes is going to be, with Web sites carrying all the details. Next month, we will be able to order whatever we want from the new lines.

CLOTHING

One thing is for sure: All the clothes are designed for durability while being beautiful and comfortable at the same time. When the five-year Parade started, the fashion elite were beside themselves because they didn't think they could "live" with the same look for half a decade. What they hadn't figured out yet was that while the clothing didn't change, you could change the accessories. Buttons, for instance, became a place to show your individuality. Soon, artists were producing unique decorative works to create a fresh feeling yet stay within the guidelines.

This accessorizing set off a huge handicraft boom around the world. And through WOW (the World Optical Web), you could order from anywhere.

Then, on the third design cycle, the real revolution occurred: programmable fabrics that allowed you to change the color and the pattern of colors on your clothes. This enabled even more chances for individual expression while keeping the same article of clothing for five years. Shortly after that, the fashion complaints stopped. The designers focused on new software designs for the clothing "hardware."

When you think back to all the waste of the last half of the twentieth century, we are managing the world a lot better these days.

For instance, thanks to a relentless effort across the world to increase energy efficiency, the world's population, which is now 7.2 billion and on trend to level off at eight billion, consumes 30 percent less energy per capita than it did in the 1980s. The major reductions came from the developed countries led by the United States of

America. The United States really got its act together starting in 2006, when a major economic study demonstrated, once and for all, that energy efficiency was truly worth its weight in gold.

TRANSPORTATION

It was known by then just how environmentally expensive automobiles were, but the cost of airplanes also became evident. Their impact on rain patterns and cloud patterns was scientifically quantified. They also impacted pollution patterns in ways that had not been understood before. All that information triggered the great train revival, led by a new design of lightweight magnetic trains (M-trains) that ran on the old tracks.

Of course, the old tracks had to be improved so that the trains could travel at two hundred miles per hour, but that cost was trivial compared to the predicted costs of more airplanes, more airports, more cars, more highways, more pollution, and all the infrastructure costs they would have involved.

The new trains are electrically powered, using superconducting magnets as the motors, which push against aluminum plates that have been fitted between the rails. They generate their electricity with fuel cells, which run on hydrogen. The power unit isn't much bigger than an old 2005 minivan's. General Motors invented the power unit. It hardly pollutes at all and is very efficient. Even the skin of the train has a special 3M coating modeled after sharkskin that reduces turbulence and makes the train cut through the air more easily.

The first M-train was built and deployed in Florida. After a tremendous fight with the governor—he came from the Bush family and oil money—the citizens passed a constitutional amendment to stimulate this new technology.

The Florida M-train ran from Tampa to Orlando to Miami for the first five years. Then they added a spur up to Jacksonville and down to Fort Myers. It made getting around Florida so easy, so inexpensive, and so fast that tourism doubled. Besides more tourism, the

M-train created more than 20,000 highly paid manufacturing and research jobs. Florida became the Northern American center of the new transportation technology. The technology spread across the nation just as fast as the tracks could be improved.

M-trains are almost silent, rolling on low-friction wheels with magnetic bearings. And because the train cars are made out of carbon fiber, their light weight does far less damage to the tracks than previous trains. Even the freight trains are much lighter now and faster—they average 150 miles per hour. For passenger trips up and down the East Coast of America, no one flies any more. In Middle America, for trips between Detroit and Chicago, Chicago and Minneapolis, Minneapolis and Denver, Denver and Salt Lake City, the train is the only way to go. The same is true up and down the West Coast and between LA and Phoenix and Las Vegas. Every route has at least two competitors, so the marketplace keeps prices in line.

Of course, Texans have their own trains moving between their six great cities. In fact, it was a Texan who solved the intracity-traveling problem: How do you finish your journey when you get off the train? Michael Dell, who revolutionized the manufacture and sales of computers, started Dell Cars and used his same manufacturing philosophy to build great-looking cars for short trips in and around the city. Dell licensed the technology from its inventor, Amory Lovins, famous for his research into energy efficiency.

He showed Dell his ideas and Dell turned those ideas into inexpensive, highly efficient, long-lasting city cars that you rent with a credit card. Almost all of the car makers in the world, except Toyota, failed to see the emerging opportunity and were never able to get much of that market. It just shows you what happens when you miss a paradigm shift. Of course, the old car makers build the trains and long-distance cars, so they are still busy, but there are a lot fewer automobiles in the world than there used to be, and it looks like it is going to stay that way.

If you buy a car today in 2030, you expect it to last for twenty-five years and go five hundred thousand miles before anything needs to be fixed. The body is foamed plastic over a carbon fiber frame. The foam absorbs enormous amounts of energy in a crash and protects pedestrians, too. And when you want to change the style of your vehicle, you simply go into a foamfab dealer and pick a new look. The old foam is melted off with electron beams. The new body is applied with foam guns and shaped and finished with lasers. The color is built into the foam. The cost: about $250 and four hours of time. The foam is recycled and used again.

Of course, people travel a lot less today than they used to. Thanks to sophisticated communications, 50 percent of the U.S. population work from their homes on any given day—which means that the home is a lot more important than it was at the turn of the century.

HOMES

Certain things are givens about homes these days. For starters, they are wonderfully energy efficient. My house, which is five years old, has aerogel insulation throughout. The six-inch walls of my home are R 50. The twelve-inch roof of my home is R 100.

My house is quiet, delightfully quiet. The windows, too, are very special. They autodim to decrease sunlight coming in during the summer. I can also dim them for privacy. They are coated with a self-cleaning material. Dirt can't stick to them, so each rain washes them crystal clear. My first-floor south wall is an aeroglas window wall, with aerogel insulation laminated between two pieces of glass. This allows light to come in but keeps heat from escaping. During the Minnesota winters, my south wall actually adds heat to my house while it is forty below zero outside by allowing infrared radiation to come through the window.

Because Minnesota is a relatively sunny state, most of my electricity and hot water heat come from my solar shingles. On cloudy days, my house gets its heating and cooling, hot water, and electricity from a fuel cell that is 4 by 4 by 10 inches, a Ballard. It uses hydrogen created with the excess electricity my solar shingles generate on sunny days. It converts that energy into electricity and hot water at about 85 percent efficiency. The hot water tank is coated with twelve inches of aerogel, so the water stays hot for seven days with no added energy.

We recycle almost all of our water. As soon as waterless toilets became the required standard for all houses, the water in the household could be filtered and reused. We have a holding tank of two thousand gallons, which is constantly cleaned and purified. But, over time, we need to top off the tank. As a result, we go through about fifteen hundred gallons of water each year. Can you believe that many households in the twentieth century went through that much water in a week?

If you walk into the kitchen, you will see the same kinds of efficiencies. First, we are a lot better at food preservation these days. Instead of freezing everything, we now can keep the flavor fresh by storing food in an irradiated, recyclable pouch. The radiation kills any bacteria inside the pouch. But, once we open a pouch, we still have to refrigerate the leftovers. Some things never change.

The refrigerator is a lot smaller with aerogel insulation. The refrigeration unit is an acoustic compressor that uses sound waves instead of a physical mechanism. It has no moving parts so it never wears out. And it uses only 15 percent as much energy as a late twentieth-century refrigerator would. The system is so efficient, you can just barely hear it hum. Because the house is so well insulated, we can cool it with the same unit that cools the refrigerator.

When you compare the cost of our house—2,500 square feet in size—to what you would have paid for a same-sized house in the 1990s, our house costs 66 percent less. Operating it costs 90 percent less. That means we have a lot more money to spend on other things.

COMMUNICATIONS

The communications matrix in our house is first class. Of course, everyone's is first class these days because it is so cheap. I have a satellite house, which means everything comes in by way of satellite.

Space satellites in the twenty-first century are mostly assemblages of picosatellites. They work in tandem with twenty to one hundred other picosats to do the comprehensive monitoring and communicating that has helped us manage the world better.

Because picosats are so small and robust, we can shoot them into space with special cannons. Developed by two physicists from Lawrence Livermore Labs in the last years of the twentieth century, they allow us to put a kilogram of material into low Earth orbit for about $20, which has allowed us to build space stations using cannon-shot building modules. And they cost 1/10,000 as much as the first international space station.

There are more than four hundred space stations orbiting now with an average of two hundred people in each station, almost all scientists and engineers whose primary work is watching our planet and watching out for it.

Since we are the only species capable of fending off a killer asteroid, we take that job very seriously. Between 2015 and 2030, we intercepted and displaced five killer asteroids whose orbits would have intersected Mother Earth within one hundred years. I marvel at how lucky humans were to have survived before we put up the orbital watchtowers.

In some neighborhoods they still use copper cable for communication because they have older houses that have been retrofitted. The new data compression technology allows the old twisted-pair, copper wires to deliver everything I get via my satellite system. We are very proud of how we have been able to make seemingly obsolete infrastructure do the work of the twenty-first century. Waste not, want not. It is true if you're willing to be clever about it.

I am a curriculum designer and specialize in simulation learning. I service the early education arena. My work is accessible through WOW and is available in five major languages plus Global. Video conferencing, collaborative teaming, virtual training—all are part of my work.

SIMULATIONS

The art of simulation has really progressed in the past fifty years. Thanks to Cray Computers back in the 1980s, the world began to understand the importance of being able to test new ideas inside a computer before testing them in the real world. Today we have simulations for almost everything.

For instance, we have gotten really good about weather simulation, which has led to much more accurate long-term forecasts. We have thrown in the towel for perfect forecasting, because chaos theory has proven that the variables are too extensive. But, we do a very good job of predicting droughts, floods, and hurricanes in a twelve-month window. Our forecasts for the next two months are increasingly accurate, which makes life much easier for farmers all over the world.

Because of our drought and flood prediction, we have, for the first time in human history, the capability of being ready with food and shelter and other essentials at the right time and place. We can't stop the floods, but we can move people to higher ground.

Now we are working worldwide to get people to live somewhere other than a flood plain. This is all part of the Earth Restoration Effort (ERE).

If there was any single effort that epitomizes the twenty-first century, it is the ERE. Early in this century, we finally did an honest assessment of the damage human beings had done to the world ecosystem. This led to the ERE.

Until the Great Accounting, no one really knew whether the planet was in better shape or worse shape than before the advent of modern science. So the United Nations convinced the rich nations of the world to spend $20 billion to find out. By 2010, accurate data had been collected. It was discovered, for instance, that the Chinese had been lying about how many fish they were catching each year— actually doubling the real number. This had led world officials to believe the fisheries were healthy. "Not so," said the new data. And that pattern of lying was found across industries worldwide.

Added to this new and correct data was new research information on how animals and plants formed ecosystems and what kept those ecosystems healthy. Then calculations were made of the value of all the ecosystems.

The damage was assessed. The long-term consequences of not fixing the damage were debated. It took five more years, but the world agreed that we must restore the living systems of the world. Today 20 percent of the world's population and 20 percent of the taxes are dedicated to that effort.

The ERE has turned out to be the first global vision for humanity. It brought nations and cultures together. And, as the forests of the world regain their health and the fisheries of the world rebound and the rare species return in greater numbers, the news is greeted with worldwide celebrations. The wilderness is returning and yet we humans are eating better than ever before and living longer. All the while, we are doing less damage to our life-giving natural support system.

Simulations have been a key part of the restoration. We are now much more careful about introducing new technologies before we take their measure. The long-term impact on our species and our planet is a required measurement. Plus, by simulating we avoid hidden costs and do substantial improvements to an idea before it becomes real.

DESIGN AND MANUFACTURE

Think about it—why waste real time and real resources when you can check out almost any idea in tech-time via simulation? And these days, with the incredible computational resources available and the vast pools of simulation expertise, it takes almost no effort and very little cost to get a good measure of an idea.

Once we have verified a design, then we can go shopping for the resources to build it. Here is where the old World Wide Web created the paradigm that still works today. We list our design/build requirements and, from around the world, proposals pour in. In a couple of days, we can put together the entire enterprise needed to manufacture, distribute, market, and sell the design. And, because we are utilizing resources already in place, it is a very efficient way of doing things. No new plants to build. No engineers to hire. No sale teams to train.

This is the New Mutualism, a new paradigm of enterprise in this century. We are finally copying Mother Nature on how she is organized. Using the strengths of difference, we fashion a kind of ecosystem to create our products. Each element contributes its special strength, and the organization generates complex solutions through the combination of differences.

Of course, the old antitrust laws had to be rewritten to recognize the power of mutualistic teams. Mutualistics became the science on which most twenty-first–century organizations are now structured.

Once that was done, the giant corporations began to disappear because they were too encumbered with inefficiencies. In their place, a mutualism of many small companies became the predominant mechanism for doing large projects.

The keys to success of these mutualisms are transparent information for the team members and extremely detailed communications links (called the Trust Protocol), so that everyone is kept in the loop. Thanks to WOW, you can build mutualisms with groups anywhere in the world as long as all follow the Trust Protocol.

With the protocol, trust is established, measured, and maintained. If a member breaks the trust, they are immediately banished from the team, and their bad behavior is reported and cataloged.

Breaking trust usually means that group is never invited to be part of a mutualism again.

The Fox manufacturing facility in Mexico is a key partner in many North American mutualisms. Consequently, it turns out more than ten thousand different products for North America on any given day.

Because the manufacturing site acts like an ecosystem, where the waste from the production of one product is another product's resource, the site is very efficient. At the end of this industrial daisy chain there is almost no waste to dispose of. Denmark pioneered the process back in the late 1900s and the Mexicans perfected it as part of their industrial leadership program started by Vicente Fox when he was president. When the products are ready, they are sent to their destinations via the M-train system.

ENERGY FROM WASTE

Here's one more comment on waste. The other great paradigm shift in this century was that waste was simply resources that we were unable to utilize. The development of the thermal depolymerization process (TDP) revolutionized the processing of waste. In a series of simple steps, it was able to take almost anything—tires, paper-pulp effluent, sewage wastes, oil-refinery residues, dioxins, old computers, plastic bottles, slaughterhouse remains—and convert it into natural gas, number 2 fuel oil, and reusable minerals. Literally anything with a carbon base could be rendered into liquid energy with an efficiency of 85 percent. Dumps, huge mountains of tires, animal wastes—all became sources of fuel and disappeared as pollution problems.

Within ten years, most of the liquid fuels of the world were being generated by TDP and, instead of energy sources being concentrated in the Middle East, energy sources became distributed across the globe adjacent to the population centers that generated waste.

This one technological revolution solved three of the biggest problems facing humans at the end of the twentieth century: impending shortage of liquid fuels; overwhelming amounts of waste to dispose of; and growing fossil CO_2 being released into the atmosphere. Now almost nothing organic is considered waste; instead, it is the feed-stock for the TDP systems around the world.

When I think back to the way we lived in the twentieth century, I shudder. If we hadn't gotten serious about operating within some thoughtful limits, we could have permanently injured the world ecosys-tems. Now, we are living better and longer, working to make the planet healthy for the next ten thousand years, and everyone across the world is better off than anyone ever dreamed possible.

3

LOCAL TECH

> *Simplicity is the peak of civilization.*
>
> JESSIE SAMPTER

Even as the debate between the Super Tech region and the Limits Tech region raged on in the early 1970s, a third TechnEcology began to take shape. E.F. Schumacher, a noted scholar, wrote a little book entitled *Small Is Beautiful.* Because he was a self-described Buddhist-Catholic-economist, he brought a different perspective to the debate. He founded what has become the Local Tech region of the future.

In the 1950s Schumacher spent time in Burma and noticed that farmers could not make good use of modern tractors from the West because the plots they farmed were small compared to farms in the United States and Europe. He also noticed that when more modern devices began to be manufactured, the older technologies disappeared until there was nothing available in the marketplace except

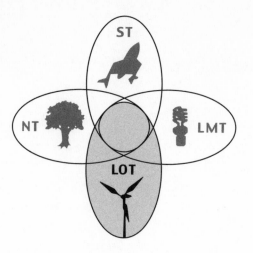

the very advanced and the very basic. So there was no technology available to fill this gap.

Schumacher coined a term to describe the technological need between advanced and basic technology—*intermediate technology*. This no-longer-available intermediate technology was much more appropriate for the farmers in Burma than the newest equipment, so Schumacher also called it *appropriate technology*. Since local needs defined what was appropriate for local use, Schumacher concluded that this appropriate technology could best be supplied from local sources.

These discoveries led Schumacher to develop the principles of what he called "Buddhist economics." They were based on the notion that *production from local resources for local needs is the most rational way of economic life.*

But the key reason that everyone should have technology appropriate to their own use is that the work everyone does is key to their satisfaction in life. If fulfillment in life is connected to work, then technology should be selected to augment that fulfillment.

When we began our cataloging of the various regions, we called this third region the Appropriate Tech region of the future. But as the five regions evolved, we realized that the term *appropriate* also connotes *correct*, and that was not our intention. Since local conditions define what is appropriate, we decided to call it the Local Tech region.

Local Techers prefer small to big, but they do not focus on size for its own sake. For example, the "village" is the model for human settlement. Villages or communities should not be too big, or they will be dehumanizing. But they should not be too small either, or they may miss the richness of variety in life. The scale is the key. Proper scale is what humans can effectively manage in any environment.

An example of proper scale is the optimum number of people working together in a manufacturing facility. Schumacher advised the Scott Bader Company, a chemical manufacturer in England, on organizing their plants and suggested keeping the plant small enough so the work would remain interesting and productive and everyone could know everyone else. Therefore plant size was set at five hundred employees. When its growth demanded more employees, it split off part of its workforce and let them establish another plant. In this way, the sense of working with people you know was preserved while the demands of the marketplace were met. In several studies done to compare efficiencies of various sizes of plants, the Scott Bader Company repeatedly placed in the very top spots even though it was, by far, the smallest company in the competition. Schumacher liked to point out that size was vastly overrated as a way to profitability and efficiency.

Like Super Techers, Local Techers are optimistic about their ability to use technology to solve problems, but like Limits Techers, they want technology to be friendly to the user and to the environment.

Guidelines

1. There is a sufficient amount of resources in the world for everyone. Sufficiency is a very important word for Local Techers.

It implies that we can define what is "enough" in the essential aspects of our lives. They believe that there are enough resources to meet the needs of humanity. We cannot all have mansions or personal space ships, but we can all have enough to live a very good life.

2. Science and technology are OK as long as they are properly scaled. Scale is very important for Local Technologists. Technology that is too big for a community to handle can get away from its citizens and become disconnected from its impact on the world. Technology that is properly scaled is friendly to the user and to the Earth.

3. Humans are the shepherds of nature. Super Techers say that it is time for us to take over evolution. Limits Techers say, "Mother Nature knows best." Local Technologists want to be the shepherds of nature. They have an active responsibility to protect the natural world and to harvest in a way that does no damage to the long-term viability of the Earth.

4. Humans need to work to become fully human. This is a popular notion today, but in 1973 it was a startling statement. Work is an essential part of your spiritual growth. Even if you have $1 billion, you need to work to become a full adult and to achieve your potential.

In the Local Tech region we will live in human-scaled villages and communities with technology that helps us to use the resources of nature in a way that protects the planet and allows us to work in a way that fully develops our humanity.

And, of course, the motto for this region is:
"Small and local is beautiful."

302 ▦
Advocates and Examples

Advocates

E.F. Schumacher. E.F. Schumacher heads the list of spokespersons who have played a significant role in defining the rules and ideology of the Local TechnEcology. He has been joined by a good number of other writers and activists to help flesh out the details of this TechnEcology.

Ernst Friedrich ("Fritz") Schumacher was born in Germany in 1911. A Rhodes scholar at Oxford in the 1930s, he moved to England before World War II to avoid living in Nazi Germany. During World War II he aided Great Britain by working with its economic and financial mobilization.

After the war, Schumacher was an economic adviser to the British Control Commission (charged with rebuilding the German economy) and the British Coal Board, one of the world's largest organizations, with eight hundred thousand employees. As a planner he predicted the rise of OPEC and the problems of nuclear power.

While an economic adviser to Burma he developed the principles of what he later called "Buddhist economics," which emphasized the importance of work in one's life and the value of local production and use of local resources.

Schumacher subsequently spread his ideas as a featured writer in the British journal *Resurgence* until his death in 1977. His landmark 1973 book, *Small Is Beautiful*, was republished in 1999.

Lester R. Brown. A leading researcher and author, Lester R. Brown has advocated a sustainable economy and given guidelines about how to build one. Brown has provided key measures of food supply, water use, land fertility, as well as a number of "vital signs" of trends that shape the future.

After a career in the U.S. Department of Agriculture, Brown in 1974 founded the Worldwatch Institute to be a research organization devoted to the analysis of global environmental issues. In 1984 Brown began editing and publishing annual *State of the World* reports. These assessments, translated into dozens of languages, have become standard references for global conditions.

Among his many influential books are *Saving the Planet: How to Shape an Environmentally Sustainable Global Economy* and *Building a Sustainable Society*.

In 2001 he founded the Earth Policy Institute and published *Eco-Economy: Building an Economy for the Earth*. The purpose of the Earth Policy Institute is to provide a vision and road map for achieving an environmentally sustainable economy and an ongoing assessment of progress in this global effort.

Hazel Henderson. An independent futurist, syndicated columnist, and consultant on sustainable development in over thirty countries, Hazel Henderson has developed a number of alternatives to established economic indicators to highlight factors important to the sustainability of the environment and the economy. She has been an advocate of alternative energy sources to replace oil and other fossil fuels.

The first version of her Country Futures Indicators™ (an alternative to gross national product) was launched in 1996 with the Calvert Group, Inc., as the Calvert-Henderson Quality-of-Life Indicators.

Henderson serves on the advisory boards of the Calvert Social Investment Fund, the Cousteau Society, the Council on Economic Priorities, and the Worldwatch Institute. She also serves on several editorial boards, including *WorldPaper* (an insert in twenty-five newspapers distributed in Asia, Latin America, China, Japan, Russia, Africa, and the Middle East), *Futures Research Quarterly*, and *Futures* (UK).

Her editorial columns are syndicated worldwide by InterPress

Services. She has published articles in over 250 journals, magazines, and newspapers. Henderson has authored several books, including *Building a Win–Win World; Creating Alternative Futures; The Politics of the Solar Age;* and *Paradigms in Progress*.

David Morris. David Morris is an author and vice president of the Institute for Local Self-Reliance, an organization that promotes humanly scaled, environmentally sound economies and institutions. He has been an advocate of the development of local economies, the co-generation of electrical power, and the creation of communities capable of becoming self-sufficient.

The author of four books and several dozen monographs, Morris is a consultant to business and government in the United States and abroad. His most recent reports include *The Carbohydrate Economy; The Mondragon Cooperative;* and *The Trade Papers*.

Donella H. Meadows. Environmental scientist and writer Donella Meadows was best known as the lead author of the 1972 book *The Limits to Growth,* which explored global trends in the population, economy, and environment.

Meadows taught ethics, journalism, and environmental studies at Dartmouth College. She founded the Sustainability Institute, which combines research in global systems with practical demonstrations of sustainable living. She also established and lived at an eco-village in Vermont known as Cobb Hill. Until she passed away suddenly in 2001 at age 59 of meningitis, Meadows wrote a regular column highlighting her insights into the world situation and how to move to a sustainable economy and culture.

Examples

We list examples of technology in categories similar to those of Super and Limits Tech. Local Tech often develops in areas left out or ignored by the first two regions of the future, but "local" (or "close to the user") is always a defining concept.

ENERGY

The choice of energy supply shapes the technological choices that each region will make in a wide variety of areas. So we will start by examining some of the major Local Tech energy preferences. Local Techers set the tone for their region by advocating alternatives to fossil fuels as energy sources.

Windmills. Where the wind blows, gather wind power. One of the local energy sources in the Local TechnEcology is wind power. Wind power currently supplies about one half of 1 percent of the electricity demand in the United States. But this supply is growing quickly, especially in the Western states.

The current leader in using wind-generated electrical power is Europe, with Denmark being at the vanguard of worldwide wind turbine design. Originally built primarily on land, wind turbine farms have proven to be more efficient on the ocean. There are wind farms off the Danish coast with others planned for Ireland, Germany, and Britain. In the late 1970s the biggest turbines could generate only about ten kilowatts each. But now the scale has increased to produce 2.5- and 3.6-megawatt turbines. These turbines are driven by rotors with a wing span wider than a 747's. Five- and ten-megawatt turbines are in the planning stage.

What is important to appreciate here is the sense of scale. A rotor that size sounds enormous for a windmill. But measured against the size and environmental impact of a coal plant, it is trivial. Also notice that you can only use wind power where the wind blows, so it cannot be a single solution to the world's energy problems.

Wave Power. Waves that crash on the shores of the oceans contain energy that can be captured and turned into electricity. The first commercial wave-powered generator to do this was installed on an island off the coast of Scotland. Called LIMPET (Land-Installed Marine-Powered Energy Transformer), the generator works by having waves roll into an eighty-one–foot-wide chamber that compresses air, which in turn drives a turbine at the top of the generator producing five hundred kilowatts or enough power to supply four hundred homes.

Norway is also working on wave power systems. And several systems are being designed to work beneath the surface of the ocean to capture the ebb and flow of both waves and tides. Where there are waves, capture their energy.

Photovoltaic (PV) panels. Where the sun shines . . . you're getting the pattern, aren't you? Photovoltaic cells, which convert photons to electrons directly, are commonly used on roofs in sunny areas. New thin-film solar panels manufactured by Uni-Solar come in rolls or shingles and can be applied directly over plywood.

There are currently hundreds of thousands of homes in the United States and millions around the world that receive at least some of their power from solar energy. Current total solar power generation accounts for only about .1 percent of global supply. However, solar-power–generating capacity is increasing rapidly around the world—by 280 megawatts (MW) in the year 2000, 340 MW in 2001, and 427 MW in 2002.

Solar Chimneys. Where the sun heats up large surface areas, the scale changes again. But solar chimney technology, again, can only be local because there aren't a lot of places in the world to support this kind of technology. One company in Australia proposes to build a solar chimney that works by collecting warm air from several square miles of glass or plastic at its base and letting the air rise through the chimney. On its way up, the air would generate electricity by spinning the blades of a number of turbine generators.

A small prototype solar chimney was built in Manzanares, Spain, in the 1980s, but to be economically viable, the Australian company estimates that their version would have to be larger than the Spanish prototype, up to three thousand feet tall or more than twice the height of the Empire State Building. Keep in mind, it is hollow, so that it requires far less materials to build than a standard skyscraper. Even though the scale of this project seems to fit Super Tech better than Local Tech, it is in actuality a Local Tech system, because it requires specific localized environmental conditions for it to work.

Run-of-the-River Hydro. Where a river runs, Local Techers will capture some of its energy, but not with the old large-scale hydro power dams. Local Techers favor a much smaller version of hydro power that can be used on a small scale. Instead of constructing large dams, the Local TechnEcology builds run-of-the-river hydro, also known as microhydro, which uses small turbines to capture energy from the water running down the river bed. No dam is required. It is a sophisticated version of traditional mills driven by paddle wheels in the stream.

HOUSING

Local Tech housing is chosen based on local community needs, conditions, and desires. The best option is dictated by the availability of local resources and weather conditions. So Local Techers advocate a number of construction materials and methods of construction.

Underground. There are many underground or earth-sheltered homes in numerous parts of the world. Weather conditions are perfect for earth homes in much of the temperate climate zones. Underground construction is also ideally suited for passive solar heating and cooling. Below-ground construction offers a constant temperature, and clever design allows natural light to brighten interior spaces.

Rammed Earth. Walls of rammed earth are constructed by erecting wooden forms that are slowly filled with layers of dirt made up of 70 percent sand and 30 percent clay. Each layer is tamped before the next is added, which results in a smooth finish as hard as stone. The surface can be plastered or painted. Almost all the building material is taken from the site itself.

Adobe. Traditional adobe construction is ancient and local. A modern substitute is machine-made blocks called pressed earth blocks, which are made onsite by an appropriately sized machine called the Village Builder: TF Sealant Block Maker, made by AECT in San Antonio, TX. They can be coated to provide additional moisture protection. Pressed earth blocks are stabilized with a small amount of cement and are produced cheaply and quickly.

Cob. *Cob* is an old English term meaning rounded mass. Cob homes are mud-walled buildings made from rounded clay-rich dirt stacked by hand, and shaped to produce smooth-surfaced walls. Cob walls are thick, up to twenty-four inches. They are also strong and durable. After drying, the cob mix takes on the hardness of sandstone. Like adobe, cob consists of mud, sand, and straw. Construction can be fun since the material can be mixed by stomping feet similar to pressing grapes.

Earth Bag Homes. Walls are made from filling burlap bags with slightly moistened clay and dirt or cement and dirt. Then they are laid on a foundation and tamped. When dry, the material hardens like rammed earth. Earth bag construction lends itself to arches and vaults and is typically covered with plaster when hardened.

Straw Bale Homes. Where straw is accessible, homes can be built of bales of straw. The bales are stacked like large bricks. Mansion Industries of the City of Industry, CA, has built several prototype straw bale homes. The straw bales used to construct walls can be made stronger than masonry. To make them, straw is spread on a conveyor, compressed into an extrusion tunnel, and heated to four hundred degrees F. The high temperature releases a natural resin in the straw that acts as a binder. The resulting straw core is covered with paper and bonded with a film adhesive. The bales have an insulation value of R 1.8 per inch and are resistant to fire.

Stack Wall. Also known as cord wood, stack wall is a modern version of log cabins with a twist. Stacked log construction uses logs cut into short pieces for easy handling. The pieces, about eighteen inches long, are stacked in the shape of the wall. Cellulose insulation is pushed in between each log and then the entire wall is covered and cemented with a special mortar. The construction is very simple and lends itself to community projects using unskilled labor.

For all of these houses, the simplicity of their construction allows them to be built with the help of friends and family. This, of course, dramatically lowers their cost, while building friendships and communities at the same time.

POPULATION

The Local TechnEcology has a very simple population premise: Don't crowd. E.F. Schumacher pointed out that as we pack people into huge, unlivable cities, there are many areas of the world that remain underpopulated. So a large global population is possible and is even a good idea to provide local variety as long as we do not try to put too many people in a small space. People need room and open space to live the good life. However, Schumacher was very careful to remind us that we must also leave sufficient space for the other species of the world, since we are their shepherds and must take care of them.

FOOD

Food grown locally (some of it with your own hands) is seen as having the greatest value for this region. The village life of the Local TechnEcology lends itself to gardens, and in this region everyone has one. Part of the rhythm of life is to participate in growing your own food. There is even the thought that locally grown food will have nutrients that are important for the local population. While there is some importation of food specialty items, Local Tech food, as much as possible, comes from the local area.

One pattern of local food growing has become a trend in many parts of the United States. Farms close to metro areas are inviting families to invest in their harvest. Several hundred families put money up in the spring to finance the planting and harvesting of all sorts of garden vegetables. This lowers the cost of financing for the farmer and guarantees him or her sold products when the vegetables are harvested. Some of the farms also require that the investing families spend one weekend day per month at the farm helping take care of the crop. This gives them a sense of ownership and also helps them understand how much work it takes to grow food.

An important food technology in this region is the greenhouse. In many climates, local food production can be maintained through using greenhouses. Fresh, locally grown food is valued by Local Techers and there is no reason that it should not be grown year-round with the help of climate-controlled greenhouses. There is little need for long-term seasonal storage because locally grown fresh food is always available in this region.

WATER

Fog Nets. An example of a local need solved with a local technology is the development of the fog net to extract moisture from the thick, foggy air. Chile solved a water supply problem with local technology developed in Denmark with Chilean scientists in partnership.

The large nets capture quantities of water sufficient for local use in arid but foggy areas of Chile.

TRANSPORTATION

The preferred modes of transportation in the Local Tech region are walking and biking.

Bicycles. Currently used in much of the world, bicycles are a Local Tech transportation solution for the future as well. They reflect the "small-and-local" mentality of this region. Bicycles also work well because villages are typically small and can be traversed quickly at bicycle speeds.

Bicycles continue to evolve as well. For example, the Think line of bicycles adds an electric motor fueled by a removable twenty-four–volt battery pack. It recharges in six hours off an ordinary household circuit. There will be continuous development of this option in the Local Tech region.

The TWIKE. A combination car and bike called the TWIKE is a three-wheeled vehicle that uses human muscle power on pedals to supplement the vehicle's electric engine. It's capable of going more than fifty mph. One charge will take you about fifty miles, and more if you pedal. The vehicle has been in use in Switzerland since 1997, and is seen as a good alternative for commuting.

Rollerblades. These wonderful devices give anyone with two good legs a chance for decent speed, exercise, and joy—all while going somewhere to do something. Village size is perfect for this mode of transportation.

Sailboats. Another favorite of the Local Tech region is sailboats. Rivers, lakes, and open oceans are all environments for sailing, both commercial and pleasure. Now the world's first hybrid wind-and-solar sailing vessel has been launched in Australia. Based in Sidney Harbor, the hundred-passenger Sydney Solar Sailor mounts solar panels that rotate to capture the sun and wind.

The Segway. Remember the two-wheeled gyroscopic riding machine that was going to revolutionize transportation in cities? You may have noticed that there are not very many of them in use. Why?

There's a very simple reason: Segways are right for a Local TechnEcology, but they are trying to operate in a Super TechnEcology. In the Super Tech world of modern America, they have no place. They are too slow for the streets (plus way too vulnerable to fatal accidents with vehicles) but too fast for sidewalks.

On the other hand, in the village paradigm of Local Tech, they would be perfect. Traffic is much more leisurely in a village than in a city, and there are fewer vehicles on the streets. Since the distances needed to travel in a village are short, going twelve miles an hour on a Segway would get you to most anyplace in the village in fifteen minutes. Of course, the users of Segways would not be kids or healthy adults—they would bike for the exercise. But for people with leg or foot injuries, or older people who don't have the strength to ride a bike, the Segway would be their chosen form of transportation.

Light Rail. E.F. Schumacher particularly liked light rail transit. Because this rail design does not carry heavy box cars, the bed for the rail can be put down much more cheaply. Because light rail is very efficient at carrying people, it reduces vehicles on highways between villages and cities. Building the passenger cars out of lightweight materials like carbon fiber decreases the amount of energy needed to move the rail cars. This technology is already in use in cities around the world.

Airships. For longer distances and larger cargos, airships are the preferred way to go in the Local Tech regions. But these will not be blimps. Blimps are simple helium-filled bags of gas. They are very sensitive to wind forces. The Local Tech airship will be a dirigible with an extremely light, carbon fiber internal frame. (There we go again, using a very sophisticated technology.) These airships will carry cargo, people, and even flying medical laboratories in the Local Tech region. And they will use very sophisticated technology: radar to watch the weather; solar cells on the top surface to power quiet electric motors; carbon fiber structure for great strength and light weight; and hydrogen gas as the lifting element rather than helium because hydrogen can be easily produced wherever there is water and electricity. The beauty of airships is that they can stop anywhere, they are relatively quiet and do not affect the flight of birds, and they require little complex infrastructure to support them.

The Zeppelin Company in Germany has reentered the airship market, and other companies in the United States and Europe are building freight haulers like the CargoLifter CL 160. At least seven companies are vying to return airships to the skies for hauling large cargo or for carrying passengers in luxury.

Again, proper scale in this region is not always small. These airships can be as long as one thousand feet. Yet, they fit the Local Tech requirements in that they do not disturb the natural world and can service a village very economically.

COMMUNICATIONS

Local Techers prefer to communicate face-to-face, but meaningful exchanges can be aided by a number of devices.

Fuel-Cell Cell Phones. Perfect for the Local Tech region is a cell phone that is not dependent on a nearby plug. Micro fuel cells will power cell phones, laptop computers, and other portable electronics. These miniature fuel cells will be refilled like cigarette lighters with ethanol (which can be made locally by fermentation) and will be no bigger than the batteries used in flashlights.

Sky Stations. Here is another place that airships can provide a wonderful support system for Local Techers. Communication devices might be networked in the Local Tech region by communication satellites suspended from airships, such as the one designed in the mid-1990s by Sky Station International of Washington, DC. Their airships will be deployed about seventy thousand feet above an area and, because they are solar powered, remain aloft for years. Since the airship is far closer to users than a satellite in outer space, it requires less power to send and receive signals and the ground equipment can utilize smaller antennas.

SANITATION

Clivus Multrum. A clivus multrum is a waterless toilet that converts household waste into compost and fertilizer right at the house. Instead of looking at kitchen and bathroom wastes as a problem, Local Techers turn it into fertilizer with the clivus. The composting system kills the pathogenic bacteria so that the materials can be used to enrich the gardens of the Local Techers. It's a local solution for the Local TechnEcology.

MANUFACTURING

Stereolithography. This is a big word to describe a big idea that can be used on a very small scale. Stereolithography works by putting down thin layers of material that build up into a three-dimensional object. The early versions made prototypes of manufactured objects so engineers could look at the design and fit. But recently, this technology has begun to create objects, using metal powders, that are the final product. Metal, plastic, and ceramic materials can be utilized in this process for making manufactured objects, one at a time. What we are moving toward is now labeled desktop or "local" manufacturing. Ideal for the Local Tech village, a computer-guided printer spray gun "prints" or layers metal, plastic, ceramic, and other materials on top of each other to build up a three-dimensional object. Small design prototypes are being manufactured now, opening the possibility of rapid local manufacture of an array of products in the future.

HEALTH/MEDICINE

In the Local Tech region, people are treated locally when possible and moved to medical centers only when needed. Individual care in small settings is the best kind of healthcare anyway. With information, diagnosis, and skill provided through the Web, treatment can be as effective in small local settings as in large medical centers.

Midwives go beyond just assisting with child delivery and are primary care attendants for women and children, while Local Tech physicians, similar to shamans from ages past, pay as much attention to the spiritual aspects of health as to physical variables.

Local Techers also embrace the use of edible medicinal plants, even though herbal plants, until recently, have been largely shunned by Western science. Plant-based medicines are an ancient practice. Archeological sites indicate that Neanderthals may have used healing herbs such as yarrow and marshmallow sixty thousand years ago. The first known compilation of medicinal plants was produced in China over five thousand years ago. Local Techers utilize plants and plant extracts from their locally grown flora to counter illness and disease. Examples are alumroot from moist, rocky banks and stream beds for inflammation as well as osha from mountain areas for stomachache and heartburn.

Even though the technology from this region rarely makes mainstream news, several countries are already exploring how to use this technology. Denmark, the current world leader in wind technology, is a prime example. Across the United States, you can find families and small communities experimenting with parts of what could be a Local TechnEcology. And several states in India are well along the pathway to this region of the future.

303 ▮▮

A Visit to Three Local TechnEcologies

What follows are e-mails written between friends who met on a trip taken during the winter break between school terms of their senior year of high school.

April 2, 2050

Dear Juanita and Susannah,

I was so lucky to meet you two on winter break! In just two weeks I came to think of you as the sisters I never had. I still haven't told Mom everything we did on break. She may guess; Mom says she was young once.

In any case, I can't wait to hear more about you and your villages and to tell you more about mine. I finished my Schumacher paper for history class when I got back to school. It made me summarize and write down many things about our life on Lake Lucy.

When my grandparents and their friends started transforming Lake Lucy in 2004, it was hard work to make the transition from the old way of doing things. But, as the saying goes, "Hard work is its own reward." And now, at the midpoint of the twenty-first century, we are on our way to creating a "small is beautiful" world. At least that's what my great-uncle David calls it.

As a graduate student in the 1970s, Uncle David met Dr. Schumacher and later enlisted to work to create the kind of world that Schumacher outlined in his book, *Small Is Beautiful*. It has turned out to be a much different world from the one we were headed for at the turn of the century. Life in my village is a result of what they did.

I take the way we live on Lake Lucy for granted, but our daily life might seem strange to someone from outside our village. I have never been to New Mexico or the state of Washington, so comparisons with Minnesota might be interesting. So, are you two interested in exchanging stories about your villages? We could each write, give details about our lives, and then see what we have in common and what is different.

Love,
Kari

April 3, 2050
Dear Kari and Susannah,

Yes, I want to learn more. I've been thinking of both of you ever since I returned home to Santa Clara.

Love,
Juanita

April 3, 2050
Dear Juanita and Kari,

Oh yes, let's do this. It will be a good excuse to stay in touch. Maybe we could write about one subject at a time so it will be easy to compare our different situations. I'd like to see the differences between your villages and mine here in Pacifica, WA. What do you think?

Love,
Susannah

April 3, 2050

Dear Susannah and Kari,

Great idea, Susannah. Why don't you go first, Kari, since you started all this? We talked a lot about clothes on break. Let's pick up where we left off.

Love,
Juanita

April 4, 2050

Dear Juanita and Susannah,

I really loved the outfits you both had on. The wonderful colors and the materials are so different from what we have in Minnesota. But, of course, we need different materials to deal with the range of temperatures we have. I don't think we talked about it, but Minnesota can go from 105 degrees in the summer to –40 degrees in the winter. I am learning to sew and knit from my grandmother, who is famous for her stocking caps.

Love,
Kari

April 4, 2050

Dear Kari and Juanita,

I am also learning to sew my own clothes. My school teaches clothing design and we study what other people in our climate zone wear in the rest of the world. But we also have a long tradition of making clothes that deal well with all of our wet weather.

I am attaching pictures of some of my shoes, since we all talked about shoes during our time together.

Love,
Susannah

April 4, 2050
Dear Kari and Susannah,

There are so many local variations in dress. I look at the styles of other places, and then I add our basic southwest print patterns to the blouses I create to make them unique to our area.

Besides clothes and fashion, I guess energy seems to influence everything about our daily life, so let's do that topic next.

Love,
Juanita

April 5, 2050
Dear Juanita and Susannah,

OK. Here goes.

Our village gets all its electrical power from run-of-the-river turbine wheels in the several streams around our village plus a larger set of river turbines on the Mississippi River, which is about twenty-five miles from our village. We are able to make the electricity we need without disturbing the flow or ecology of the river.

We also supplement our river electricity with solar panels on our roofs because, surprise, Minnesota has many sunny days. To help heat our house in the winter, we have a south-facing greenhouse that traps heat during the day. At night it is fun to walk on the tiles because they are so warm.

How about you?

Love,
Kari

April 6, 2050

Dear Susannah and Kari,

Since we have very few cloudy days here in New Mexico, our locally manufactured silicon shingles on the roof plus our community wind turbines fulfill all our power needs. Every home in the village has solar cells on the roof, so every home is energy sufficient for its basic needs. For peak power needs, we all draw off the village wind turbine farm.

One more energy thing—I know this isn't about my village, but my cousin Hanna lives in the Four Corners area and they use several types of solar energy. Hanna is an engineer and she works on the power tower. If you have never seen a power tower, it is amazing. A bunch of mirrors all focus the sun at the tower and that creates a lot of heat. They refine ores and produce specialty metals that are very pure. They are careful not to interfere with the local ecosystem.

Love,
Juanita

April 6, 2050

Dear Juanita and Kari,

Here on the Pacific Northwest coast, our natural energy source is wave power. Pacifica is on the ocean and we were among the first to exploit the rolling wave action of the ocean to generate electricity. So our ocean wave technology is very advanced and our local generating plant supplies all the electricity we need. This is fortunate because our skies are generally cloudy, which means we can't use any solar power options.

After hearing about the alternative energy sources that we all use, I am curious about our housing options. Let me start by describing my home, and then you two chime in.

Here in Washington, wood is plentiful so it is our chief construction material. Our home is a stacked log structure that we built. I say "we," because our house-raising was a community project, where everyone in our neighborhood helped out. My dad and I cut the logs into eighteen-inch pieces and our neighbors stacked them, pushed some cellulose insulation in between each log, and then chinked them with special elastic mortar. Dad says it should stay tight for at least one hundred years. These homes are very pretty because you see the circles of wood with the soft tan between the circles. I have attached a picture of my house. That's me with my dog, Goofy.

Love,
Susannah

April 7, 2050

Susannah and Juanita,

We built our house, too. Our extended family and another family who are good friends of ours got together to design and build it. Everyone helped—even my little brother, Kyle. (I'll attach a picture.)

The construction didn't take long. It is a straw bale house and we used modular construction materials so we could do all of the assembly work ourselves. My dad told me that by building our house this way, we saved a lot of money and that allowed us to build a much bigger house than we otherwise could have afforded. The three-thousand–square-foot house is all on one level.

Since we have two families living in the house, each has its own wing. There is a twelve-by-fourteen–foot bedroom for each person with the shared dining area, kitchen, and family rooms in the center. This is all we need for the eight people in our three-generation family plus the four in Justin's family. They are such good friends and we get to see each other often but also have private space. Having the help of this large networked family also makes work go faster.

Our house has a greenhouse on the south side. The inside wall is loaded with special containers where we grow vegetables. We have two fish ponds, one of them inside the greenhouse. The pond provides humidity for our home in the winter and the greenhouse keeps the pond warm even when it is very cold outside.

The pond is 10 feet by 15 feet by 4 feet deep and provides a breeding ground for about two hundred fish. We harvest about 15 percent each month. Every day we drop our waste greens into the pond for the fish to eat. Our other fish pond is outside. It is the same size as the indoor pond and works just fine in the summer. It freezes every winter, so we have to harvest the fish by ice fishing. Only Minnesotans would think it was fun catching fish in the winter. It's probably a genetic thing. We think this makes the fish taste better (which may also be genetic).

Of course, we recycle everything including all household wastes. Do you guys use a clivus multrum? It's a waterless toilet that converts all the household wastes into compost. We collect the compost four times a year to fertilize the gardens. I still cannot believe that our ancestors wasted all this fertilizer back in the twentieth century and washed it all away by using water in their toilets!

Love,

Kari

P.S.: I have attached a picture of my house.

April 7, 2050

Susannah and Kari,

Kari, thanks for the picture of your new house. The straw bale construction looks quite solid and so much different from how we build houses here. The trees, vines, garden, green-house, and herb wall make everything green all over. That is such a contrast to my village, where, even though we have enough water, the dominant color is brown. I've included a picture of my house to show you what I mean.

Our house was built of pressed earth adobe by my grandfather and his brother in 2020, using one of those little adobe-making machines. My father enlarged the house in 2030 because our extended family is large. My parents, four grandparents, two brothers, and I live in our wing of the house. Mia (whose mother and father are my godparents) and her family live in the second wing. There are seventeen of us all together. Many hands make the work lighter! It is built so that it does not seem crowded at all. In fact, our village has become known for our building designs—the painted adobe walls of homes in our village are considered works of art throughout the area.

We also tend our garden every morning and have our fish pond in the greenhouse. We do not have a fish pond outside because the water would evaporate too quickly. We grow about 50 percent of our food and get the rest from partnership farms near Santa Clara. I don't know if you guys have partnership farms. The way it works is we and one hundred other families make an advanced payment to the farmer in the spring. We all fill out a survey saying what we want her to grow. She then plants and tends our crops. Each family has to go to the farm for two days once a month, which gives her harvesting help and keeps her costs down. We love to go because we meet other kids our age and get to see our food growing in the fields. Along with the beans, corn, yams, and other vegetables, we have an ample supply of beef from the range cattle that live in our part of the country.

You are lucky in Minnesota to have so much running water. We have mastered water conservation here in New Mexico. We use fog nets to capture extra water from the early morning air.

We take care not to let our water technology interfere with the local ecology. Grandfather says we should not change the surrounding environment because it supports all the species that live here. He was trained as an environmental biologist and still gives lectures at school about how to protect and preserve our local ecology.

Love,

Juanita

April 10, 2050

Juanita and Kari,

Thanks for the information about your food. Since I just had dinner, I'll add a little about our food and how it overlaps with our healthcare.

We like to get up early every morning and tend our garden before breakfast. We also grow about half of what we eat, either outside in our garden or in the greenhouse. The special growing areas that I tend are the herb wall (where we use hydroponics) and the gourd section (where we grow the plants in the dirt). I've learned a lot about herbs and spices. My aunt is teaching me how to use the herbs to help with health issues. The gourds are very interesting. The shapes and sizes that we can grow in this part of the country are so much fun! I have grown several varieties that are so big that I have to get one of my brothers to help pick them.

Our medical records show that our health is better when we eat food grown in the soil near where we live. We also only eat the fruits of the Earth in the season thereof. That's the way my grandfather says it. Since we have four distinct seasons in this part of the country, we have great variety in our diet by season. We grow apples, potatoes, and fish of several different kinds. Since we are on the ocean, our fish farms are a major part of our life as well as producing a major part of our diet.

Working in the garden is also good exercise and gives me a good appetite.

Love,
Susannah

April 11, 2050
Dear Juanita and Susannah,

Our local conditions have also influenced what food we have here in Minnesota. For as long as anyone can remember, we have grown grains and cereals such as corn and wheat and soybeans. Dairy products are also very important, with many farms around our village producing milk, cheese, and butter. We also raise cattle, chickens, turkeys, and pigs for protein.

Love,
Kari

April 11, 2050

Dear Susannah and Kari,

My grandmother has been ill recently and she has been well taken care of by our nurses. They are helping her by taking her for walks and Grandmother talks about getting back to her work at the elementary school. She is also in a quilting group and has been missed during her illness because she is one of the chief designers (she has tried to teach me how to design, without success so far). If she needs additional care, the area hospital is a few villages away and we could get her there quickly on the light rail.

Because of my grandmother and her friends, our quilt designs are so well known in the region that they are simply called Santa Claras. Our clothing designs are also done in our own style. I can't wait to hear from you two.

Love,

Juanita

April 12, 2050

Dear Juanita and Susannah,

Health and work are also connected for us in Minnesota.

After breakfast, Mom rides her bike to work at the egg factory, which is located a few miles down the road from our house. She loves to go to work every day because she owns the factory.

Of course, so does everyone else who works there. That's how companies are organized in our area these days. Everyone who works for the company owns stock in the corporation. When employees leave the factory, they sell their stock back to the company and it is issued to a new worker.

There are sixty-six employees in the factory. It was spun off from the egg factory in Breckenridge, which had reached its limit of five hundred employees. I am sure you have discovered this same size limit for organizations. My dad tells me that E.F. Schumacher himself figured out the five-hundred–person limit.

Mom's job is to organize the work so that it is challenging and satisfying to the employees. She says that managers used to think that was not the most efficient way to organize work, but Mom has found that in the long run productivity is increased when people like their jobs. It is hard to believe that in the past people found their jobs boring or distasteful. Mom makes sure that this is not the case now. It is very difficult for people to be happy in life if the work they do is not satisfying.

Among our village exports are chickens and eggs, which we are able to trade with other villages for the things we do not make ourselves.

Anyway, everyone is leaving for the day, so this is all for now. Hope all is well with you.

Love,

Kari

April 12, 2050

Dear Kari and Juanita,

My mom and dad both work in the village assembly miniplant, which provides our major exports. Father coordinates the deliveries by airship of parts from many different villages. We assemble the products and send them, again by airship, to markets all over the region. Mother is the marketing manager for our off-road bicycle products and she told me we sell one, called the Prairie Schooner, to your village, Kari.

Our fuel-cell–powered bicycles are our village's most successful products. The combination of clean energy, light weight, and dependability makes them really popular. Mom says the extrawide tires and comfortable saddle make the Prairie Schooner perfect for conditions in your part of the world. Pretty soon the bike plant will reach five hundred people working for it and then maybe they'll associate with you guys in Minnesota and help start a plant up there.

Love,
Susannah

April 15, 2050
Dear Susannah and Juanita,

You both were writing about your parents' work. We also believe part of our work is service to the community.

Dad is on the tree-and-shrub committee of the village. They spend their community service time planting and growing things everywhere. They make our village quite beautiful, and Dad says that since we are on the edge of the prairie, the plantings also prevent the soil from being blown away.

Everyone in our village volunteers ten hours per month to serve on committees such as this one. It helps to make everyone feel part of the community. In the old days, people used to pay taxes to hire other people to take care of their villages. Can you believe that? Our village has ten thousand people. So that means we have one hundred thousand hours per month to do the village chores. And everyone down to six-year-olds helps out.

I work with the little kids, teaching them how to ride their bikes. We have a special park where they get together every afternoon. My older brother takes care of the computers to make sure they are updated and virus-free. And we all get to know one another better. Do you do the same thing in your villages?

Love,
Kari

April 16, 2050
Dear Kari and Susannah,

Yes, Kari, we have a volunteer system here in Santa Clara, but ours is different. My brother does the same thing you do. I like to help in the hospital. Someday I want to be a healer. I have just begun my tutorial in herbal medicines. Before I am through my five-year training, I will have memorized the five hundred most important medicinal uses of plants native to our great southwest region.

For example, I have learned the proper dose of alum root found in the rocky banks of our stream beds for treating mild inflammation and osha from the mountain to treat stomachache.

People can volunteer for the same public service jobs only three years out of every six. This is true of positions ranging from public administrators to police officers, each filled by people volunteering ten hours per week. This means that several people hold the same position, like our mayor, much as when ancient Rome had two consuls who shared power.

Even Don Poncho Sanza, who is one of our greatest leaders, can only volunteer to be mayor three years at a time. Then for three years he does some other community service job (although it is true that Poncho has served as mayor of Santa Clara on four different occasions). Like all villages in our region, no one here is tempted with too much power for too long a time. Dad says someone named Lord Acton had it right: "All power corrupts and absolute power corrupts absolutely." We get things done because of our sense of community service, which is also a major theme of our education system.

In fact, I have schoolwork to do so I'll sign off for now. The three of us seem to study many of the same things. Since we all have access to the same global databases, I guess that is not surprising.

Love,
Juanita

April 17, 2050

Dear Juanita and Kari,

Thanks for your letters. It is so nice to hear from my spirit sisters. I just finished my intermediate technology class. This class dealt with one of my favorite Schumacher ideas.

My team worked on a wave power project. We tracked the history of this technology. Next week we are going to spend several days at one of the older plants and then go to the newest to compare the two. It should be very interesting.

Sometimes I think I want to grow up to become a Local Tech engineer. There are so many fascinating problems to work on.

I really love my school and my teachers. That's why my community service will be in education for the next three months. I will be helping out with the six-through-eight–year-olds.

And, for my next community service project I am going to try to work in open space management.

Love,
Susannah

April 18, 2050

Dear Susannah and Juanita,

Every day after school, I put in my two hours of work at the co-op grocery store. I learn about food and nutrition by doing this and it is part of my schoolwork. Botany and nutrition are great interests of mine. Then, I bring home some of the food for dinner. Tonight, Uncle David is taking his turn at cooking, although he is not quite as good as Grandma.

I agree it's good that we have learned some of the same things in our schools. It makes it much easier to relate to people in other parts of the world. We are perfect examples of how well this works. The schools and our families are the centers of our life. By the way, do either of you have pets?

Our families have many pets. My favorite is Buffy, our blue-eyed red-and-white Siberian husky. We are mostly dog people, but we also have a ferret and six geckos that help with insects in the greenhouse. We take care of our pets and they take care of us.

In fact, I'm going to play with Buffy right now. We like to run and play ball. When I talk to her, Buffy seems to understand me. Dogs are very smart and Buffy seems to sense my every mood. Learning to relate to other living things is an important part of our education.

Love,

Kari

April 19, 2050

Dear Juanita and Kari,

My grandpa was remembering at dinner last night an old

friend who played baseball back when we had professional sports. His friend Jim played for the Seattle Mariners. Of course, I can't imagine anyone playing sports for money and no one cares any more because all the professional teams have folded. Grandpa told me that what ruined it was the players' demanding millions of dollars just to play a game. I think he was teasing me. By the way, I am the scrum-half on our women's rugby team.

My favorite community activity is drumming. When everyone joins in, the feeling you get is like nothing else. It is always amazing to me how impressive the sound is when everyone starts playing different drums with each going his or her own way without direction. But it comes together so well in the end. I guess it is another good example of self-organizing systems that we learn about in school.

After this e-mail is done, Dad and I plan to have a long talk. I like to spend at least an hour a day talking with someone I love.

Love,
Susannah

April 21, 2050

Dear Susannah and Juanita,

I have a big bass drum that I play. I hope we will get to drum together some time. By the way, do you guys have sister villages? Lake Lucy has sister village arrangements in nine different ecosystems around the world, like Virginia Water in England, Monrovia in Liberia, and Gorasin in Bangladesh.

We have student exchanges with each of these nine villages and the cooperation we have had on projects has been fantastic, much like a drumming session coming together. We should start a cooperative with your villages as well.

We communicate with these villages worldwide through the Econet. Our village was one of the first to support the Sky Station dirigible project. I love to look up at it at night. I love its soft glow. Dad told me that it hovers up there at about one hundred kilometers and they only have to bring it down once a year to do maintenance things. Pretty amazing, huh? I wonder what the world looks like from up there.

Love,

Kari

April 23, 2050

Dear Kari and Juanita,

I enjoyed what you said about what you do in your villages for entertainment. We have been drumming here forever and I like it very much. We like to dance, sing, and tell stories. My brother Michael is a great storyteller. This is the kind of entertainment I love the most, but I do want to try kite-surfing when we go to Hawaii.

In Pacifica we have sister villages, too! It helps us feel part of the larger world community. I agree we should try to get our villages to form a sister cooperative.

I noticed one of the villages in your co-op network is Gorasin in Bangladesh. Gorasin is in our co-op as well! They are an inspiration to us all. Did you know they invented "new agriculture"? It seems so obvious now that agriculture can only work well if you have local control of agriculture based on local expertise. But, back in the 1990s everyone assumed the global experts knew everything. Of course, today our expertise is based on shared databases that collect information from around the world, but we still know best how to utilize that knowledge on the local level.

Love,
Susannah

April 23, 2050

Dear Kari and Susannah,

This weekend Mom and I are going to take the Silver Liner airship to Kansas City. It will be a very relaxing trip. We are going to the regional light rail transit hearing. We hope that very soon our community will be included in the regional light rail transportation system. In fact, since I have done a couple of school projects on the subject, I will be one of the people testifying at the hearing.

It will be interesting to see Kansas City during the hearing because Mom says it used to be a large city. She says the villages of the Kansas City area were among the first to decentralize and set up the "area cooperative" structure. Now there are more than three hundred official villages but none of them has over ten thousand people.

We have much to learn from area cooperatives like Kansas City, so Mom and I are lucky to be able to travel there. We're very excited about it.

The trip will also give us a chance to talk about "mother–daughter" things. As I get older, these talks are among the most rewarding and important that I have. Besides, I love to travel by airship. It is so quiet and you can see the land below so well.

That's all for now. I am going to take a dip in the community solar-heated hot tub. I can't wait to hear from you again!

Love,
Juanita

April 24, 2050

Dear Kari and Juanita,

Juanita, I am excited to hear that you and your mom are taking an airship to Kansas City. I have taken an airship, a Boeing 937, to the Seattle village complex and my parents have taken the larger 957 all the way to Hawaii.

When traveling locally we like to use Rollerblades. We can work our muscles and enjoy the great outdoors in the Pacific Northwest. They are wonderfully convenient for village-scale travel and they're fun too. Older people or those in a cast are more likely to use the Segway to wheel around town. But when they can, people generally would rather provide their own power most of the time. Of course, bicycles are a mainstay for traveling around here.

What kinds of bikes do your families have? We've just gotten our bikes upgraded with new fuel cells designed and built right here in our village. I can average almost twenty miles an hour with the help of the electric motor. I only use mine for climbing big hills or finishing long rides. But it sure is nice to have a little help when I'm really tired.

Love,
Susannah

April 24, 2050

Dear Susannah and Juanita,

We have similar bicycles to yours—some powered, some not. The land is pretty flat here so we can usually pedal. And ditto on the Rollerblades and Segways, which are very popular when you have to carry something heavy. People my age especially like to use Rollerblades because as you ride you can check everyone else out. Don't laugh—I know you both do it too, especially when boys are around.

You know what we haven't written about? Manufacturing! For small objects we employ desktop manufacturing, although some of our older folks still call it stereolithography or 3D lithography. It's official name is the Replicator but most of the young people call it the "Rep."

All the families in each neighborhood share an N 1000 Replicator for objects smaller than one half cubic meter. For larger items, we can rent a two-meter system at the village tool shed. We buy the software and the Replicator sprays the object onto the manufacturing pad a layer at a time by mixing the appropriate raw materials. We rep most of the things we need. Do you use Replicators as well?

Love,
Kari

April 24, 2050
Dear Kari and Juanita,

Kari, yes we do. Grandfather says the best thing about the Replicators is that no one has to carry an inventory or keep a supply of old products. We can also order software for new designs and "replicate" them right here in our village. Of course, we order larger manufactured objects through the assembly plant or have them delivered by airship.

Love,
Susannah

April 25, 2050

Dear Susannah and Kari,

I guess the Replicator and bikes are two of the tools we all seem to use.

I am going to a dance tonight and I'm making a dress especially for the occasion. So, I have to finish it this afternoon. In our village we often make our own clothes. We enjoy designing and picking our own colors. We can also ensure that each piece of clothing is appropriate for the weather and season. Our clothes in Santa Clara are as much an expression of our local identity as our quilt patterns and wall decorations. I know it's the same for you two.

The dance is our annual Viva Zapata dance. We are reviving some twentieth-century dances and I'm going to do a dance my grandmother taught me called the macarena.

I wish you could both be here!

Love,
Juanita

4 ▬▬▬▬▬▬▬▬▬

NATURE TECH

401 ▰
Overview and Guidelines

> *For every problem, Nature has a solution.*
>
> ADM CORPORATION

The first three regions of the future are full of artifacts—tools crafted directly from the hands and brains of humans. The fourth region of the future, Nature Tech, is also full of tools, but almost none of them are human-made. They come from Nature herself.

Prehistoric Nature Techers used yeast to raise bread dough and bacteria to make cheese, yogurt, and beer. They took their herd animals, picked the best, and mated them to produce stronger, more productive animals for food, labor, and protection. In the same, slow way, they also expanded and improved their apples, their maize, their wheat, their potatoes, and their mangos.

Today the Nature TechnEcology has been changed fundamentally because humans have developed tools that allow direct manipulation of the DNA of organisms. Using the techniques of gene splicing and recombinant DNA technology that have been developed

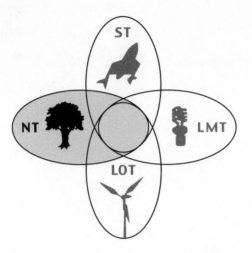

in the past twenty years, we can now combine the genetic elements of living cells that would have never been combined in Nature. And we now operate in tech-time instead of natural time. This new ability recalebrates the fourth region of the future by speeding up innovations and gaining molecular control of those innovations.

The Nature Tech region is based on the assumptions that humans can use the biological elements already operating on the planet in new ways and that we can also take their DNA and thoughtfully adapt it in such a way as to benefit the human species without disrupting the natural world.

Like all other Techers, the Nature Techers are very optimistic about their ability to use technology to solve problems. But, as their starting point, they look to the natural world for solutions that have been generated over the past three billion years before trying to invent anything on their own.

Already, however, there is a growing split within the Nature Tech practitioners. Some want to use only what Nature has already developed, with no additional human manipulation. They believe her solutions do not require any augmentation. Other Nature Techers are not afraid to utilize their new biotech tools to create novel combinations of Nature's solutions. For instance, we already have changed bacterial cells, making them capable of producing human hormones. We have stimulated cows to produce more milk from the same amount of feed. We have increased a plant's ability to absorb toxic minerals from the soil.

How these differences will work themselves out has yet to be determined. It may cause a schism in the Nature TechnEcology of profound proportions. But, as of now, here are the assumptions and value statements that apply to all of Nature Tech.

Guidelines

1. *All human needs can be fulfilled using Nature's systems.* Issues of food, issues of medicine, issues of energy, and issues of housing can all be solved using the biological resources that exist on this planet.

2. *Mother Nature has already solved all of our problems. Science's job is to find her answers and create compatible technologies using those solutions.* Nature Techers believe, with good justification, that with three billion years of experimentation, Mother Nature has addressed and solved problems similar to what we are facing today. If we look hard enough, we can find her solutions and adapt them to our specific needs. This guideline places biology at the top of the pecking order in science. Physics and chemistry serve the needs of Queen Biology.

3. *Our relationship with Nature is that of a co-equal partner.* Again, this relationship with Nature is different from relationships with Nature in the first three regions. We are not shepherds of Nature, we don't defer to Nature, and we don't leave her behind. Instead, we are equal partners with Nature, working with her to the benefit of all living things.

4. Our work is to learn to live well with Nature. We have been fighting with her for ten thousand years. No more should we see Nature as an enemy. From now on, we will develop technology that creates a mutualistic relationship between humans and the living systems that cover the surface of our planet.

Our goal is to fully integrate with Nature and use the knowledge of biology to co-direct the ongoing evolution of the natural order. All will be in harmony with the natural circle of life.

The slogan for this region is:
"Nature is beautiful."

402

Advocates and Examples

Advocates

The advocates of Nature Tech are much less well known than the advocates for other TechnEcologies. That is because this region has really been camouflaged by the explosive birth of biotechnology. Five years ago when we would speak about this region, we could only give primitive examples of Nature Tech. But the pace has quickened to the point where almost every day there is a new announcement about a Nature Tech solution. The technology, in both its natural and adapted states, is growing increasingly sophisticated.

James Watson and Francis Crick. Watson and Crick are credited with deciphering the double-helix structure of DNA in 1954 and defining its role in heredity. This new understanding of DNA triggered the birth of the biotechnological revolution that is reshaping our understanding of the natural world and our capabilities of altering natural processes for our own benefit.

Maurice Wilkins also played a crucial role in the DNA discovery and, as a result, shared the 1962 Nobel Prize with Watson and Crick. It is now recognized that the discovery would not have been possible without the brilliant work of Rosalind Franklin, who died at age thirty-nine. Both Watson and Crick were at Cambridge University in England while Wilkins and Franklin were doing their research at King's College in London.

James Lovelock and Lyn Margulis. The biophysicist Lovelock and the microbiologist Margulis are authors of the idea that Gaia (a mythic name for the Earth) acts like a living organism to create a stable environment for all life. The individual living organisms that make up the 106 different ecosystems so far discovered are, in a sense, the organs, muscles, and nerve tissue of the largest living system—Planet Earth.

In his book, *The Ages of Gaia,* Lovelock looks at the macro level of the Gaian system. He argues that interacting physical, chemical, and biological processes may control such things as the level of oxygen in the atmosphere, the formation of clouds, and the saltiness of the oceans, which in turn make the planet habitable for a wide range of living organisms.

That our planet is a "superorganism" and that evolution is the result of cooperative, not competitive, processes have become the two fundamental components of the Gaian theory that Lovelock and Margulis developed. The implications of the theory are profound because it means that it is no longer sufficient to say that organisms that leave the most progeny will succeed. It becomes necessary to add that they can do so only as long as they do not adversely affect the environment.

Margulis's research focused on how single-celled organisms play a critical role in the regulation of positive planetary conditions and how cooperation, not competition, has been the guiding principle for the development of complex life forms.

Michael Rothschild. In *Bionomics: Economy as Ecosystem,* Michael Rothschild takes the principles of Nature as illuminated in the science of ecology and shows how human economics act in much the same way as an ecosystem does. He goes on to show that our greatest economic errors occur when we violate ecological principles.

William McDonough and Michael Braungart. McDonough and Braungart wrote *Cradle to Cradle,* a treatise on how human beings can interact with Nature without destroying her. McDonough is an award-winning architect who has focused his designs on the use of natural materials to integrate his buildings with Nature as completely as possible. He has designed for Ford Motor Company an assembly plant whose roof is covered with growing plants. This helps the building stay cool and captures and purifies rain water. It is built from materials that can be returned to the earth without damaging it. His constant theme is to emulate the cyclical pattern of use and reuse or, as he calls it, intelligent decomposition.

Janine M. Benyus. With *Biomimicry: Innovation Inspired by Nature,* Janine M. Benyus became perhaps the most important author for Nature Tech. Ms. Benyus popularized the term *biomimicry,* which is the process of copying and adapting solutions developed by living organisms to solve our own problems. She speaks of the dangers of losing the biodiversity of the planet and, with it, the profound knowledge that each living organism has accumulated as the result of its successful evolution.

Examples

Being a product of Nature itself, this technology has a much different feel from the technologies of the first three regions.

ENERGY

Hydrogen. Energy is always an issue for human civilization. Of course, coal and oil are "natural" fuels, but they carry a penalty of adding more fossil CO_2 to the atmosphere. Hydrogen is a much better fuel, because it becomes H_2O when burned. But, can it be produced naturally? The answer, according to researchers at the University of California at Berkeley and the National Renewable Energy Laboratory in Golden, Colorado, is yes. They have found that the green algae *chlamydomonas reinhardtii* can be stimulated to produce hydrogen instead of oxygen.

Lead researcher Tasios Melis believes that the plants are producing less than 15 percent of their true capacity. He is now focusing on how to improve the yield by manipulating their DNA.

There is also research being done on bacteria that do not require a water environment to produce hydrogen, but instead break down organic wastes directly into hydrogen.

Ethanol. Ethanol made from agricultural wastes is the focus of Iogen, a company in Ottawa, Ontario. They are using enzymes made from a fungus that has been "engineered" to convert cellulose in corn waste to sugars. The sugars are then fermented to make ethanol. If all the corn waste in the United States were processed this way, it could generate ninety-five billion liters of fuel each year.

Electricity. Even as work is being done on hydrogen production, researchers at the University of the West of England in Bristol are developing microbial fuel cells. These living devices take in fuel (in the prototype, sugar) and convert it directly to electricity. A forty-watt bulb can be kept running for eight hours on about fifty grams of sugar. Just like their American counterparts, the English researchers are working on expanding the diet of the bacteria to convert waste products and improving their efficiency with DNA alterations.

These two research efforts represent developing Nature Tech solutions for transportable energy and electricity. But, as Nature Tech becomes more sophisticated, we may find out that we can meet our energy needs in ways that are not now obvious.

Home Heating. What is the best way to heat a house? If we assume that we're dealing with a well-insulated home, then the Nature Techers have a fascinating proposal. We might apply a discovery made by Roger Seymour at the University of Adelaide in Australia: the "furnace" plant, *Philodendron selloum*. This flowering plant is exothermic. That means it produces excess heat much like an animal. In fact, one blossom releases as much heat as a three-kilogram cat. So, inside your home, you would not have a furnace to keep you warm. Instead, you would have a bank of plants "programmed" to flower during the cold season and they would heat your home.

This is how Nature Tech works. Find one of Mother Nature's creatures doing something you need to have done. Then optimize it using biotechnology.

MATERIALS

Fibrous material is a natural for this region. But Nature Techers are engineering the natural processes to meet human needs more directly.

Colored Cotton. One example is colored cotton. Researcher and entrepreneur Sally Fox has been crossbreeding cotton plants to create varieties of naturally colored fiber that can be used to weave dye-free fabric. Of course, by eliminating the need to dye the cotton, she also eliminates many processes that pollute the environment, another advantage of Nature Technology.

Spider Silk. Another natural fiber, never before accessible for human use, is being commercialized using the Nature Tech principle of partnership. Nexia Biotechnologies near Montreal, Canada, has transferred the spider-silk–making genes into mammalian cells. These cells secrete a pair of proteins that are then extruded through tiny holes. The process triggers the proteins to self-assemble into continuous fibers that are lighter yet tougher than Kevlar (a Super Tech material) and nearly as elastic as nylon. Though this fiber is not as flexible as dragline spider silk, it is better than anything man-made.

Because of its strength and low weight, spider silk has a tremendous range of applications from parachutes to replacement tendons and ligaments, not to mention becoming a trendy fashion fiber.

Plastics. Researchers at Cargill Dow Polymers have developed a natural plastic called polylactide polymer (PLA). The basic ingredients start with corn as a feedstock. This process is as much as 50 percent more energy efficient than Super Tech processes for making conventional plastic resins.

But Nature Techers are moving rapidly up the ladder of sophistication. Instead of making biodegradable plastic in large, old-tech factories, scientists are getting the plants to make it all by themselves. Research begun in 1994 by Chris Somerville and his team at Stanford resulted in a process that gets plants to produce polyhydroxybutyrate (PHB), a biodegradable plastic. Twenty percent of their dried single-celled plants consists of PHB. While this version of plastic has only limited applications, the research points the way to producing a much wider range of useful polymers the "natural" way.

Glue. The Idaho National Engineering & Environmental Laboratory has analyzed the sticky stuff mussels produce to anchor themselves to rocks, pipes, and boat hulls. Their epoxylike substance works underwater and is more tenacious than SuperGlue. The next step is to mimic its chemical makeup and find a container that doesn't get glued shut.

Coral. Coral reefs are wonderfully engineered natural structures. The Australian Barrier Reef is the largest structure in the world assembled by living things and has withstood hurricane forces for millennia. Research sponsored by the United Nations and led by Thomas Goreau and Wolf Hilbertz of the Global Coral Reef Alliance have discovered and prototyped a way to grow coral material on command. It turns out that running small amounts of electricity through a metal grid triggers the precipitation on the metal of a kind of limestone made up of calcium carbonate very much like coral. They have named the material "Biorock™."

This work offers a way to rejuvenate reefs and the data suggests that the Biorock™ reefs are actually more capable of withstanding coral bleaching than indigenous coral. Their most recent experiments have been in Bali with a three-hundred–meter stretch of artificial reefs. Also, using the Biorock™ process, it may be possible to grow structures like pillars, panels, and tubes in the ocean. These ocean-manufactured elements could form the basis of a low-cost housing and infrastructure industry for countries with seacoasts.

COMMUNICATIONS

Fiber Optics. Humans invented optical fibers in 1951 and began to apply them in a practical way in the 1990s. Certain sponges did the same thing over one hundred million years ago. The Rossella sponge (*racovitzae*) flourishes in the Ross Sea in Antarctica. It is covered with "spicules," long, thin, sharp needlelike structures made of silica. They are capped with star-shaped forms previously thought to be part of a defense mechanism.

New research has discovered that the spines are actually optical fibers that capture and transport light into the interior of the sponge. The star-shaped structures turn out to be light-gathering devices to bring in more photons to the fiber optics channel. The purpose is not yet proven, but scientists suggest that the diatoms living in the belly of the sponge make use of the light and pay back the sponge with nutrients. The design and performance of the sponge's optical fibers are every bit as good as human design with one added advantage: The sponge's technology is immune to the damages of water. Human-designed optical fiber is destroyed by water.

Microlenses. While the Rossella sponge was perfecting fiber optics, the brittlestar, a relative of the starfish, was perfecting lenses. It has evolved a series of microscopic lenses that are nearly perfect optically. Why is this connected with fiber optics? Because human design for fiber optics calls for microlenses to control the signals moving through the optical fibers. And it is clear that the brittlestar's lenses are dramatically superior to our own plastic versions.

These last two examples are two extremely sophisticated technologies developed by Mother Nature over eons of time. Our work is to understand and apply her knowledge to our communications problems.

It is becoming clearer and clearer that we have vastly underestimated the level of communication going on in the natural world and the variety of modalities.

Single-Cell Chatter. For instance, bacteria seem capable of communicating extraordinarily precise information to one another through the air. It had been shown in earlier research that bacteria have a very sophisticated communications system in liquid media. Biofilms—large, complex arrays of bacteria—are masters of this medium.

But a report in *New Scientist* magazine on research done by Richard Heal and Alan Parsons of QinetiQ Research Organization indicates that bacteria may also communicate through the air. Though it is only over very short distances (according to our standards), their experiment suggests that one group of bacteria can tell another group of bacteria to "turn on" their antibiotic resistance genes and thus be protected from an array of antibiotics. The result of these communications allowed the bacteria exposed to various antibiotics not just to survive, but to multiply. This kind of airborne communication may explain the rapid development of resistant bacteria in hospitals.

Tree Talk. Larger plants also have a complex communications system. Twenty years of research by Jack Schultz of the Max Planck Institute for Chemical Ecology have produced evidence demonstrating that trees (being attacked by insects, for instance) can signal distress to their neighbors with airborne chemicals. Those signals include instructions for how to defend against the attackers. Schultz observed that within forty-eight hours, the trees receiving the signals began to produce compounds within their leaves that reduced the insect activity by more than 90 percent.

Consider the implications of the level of this communication. What other kinds of sophisticated conversations are going on in media that we don't have a clue about? This represents a paradigm shift of profound proportions. What if we could learn the language? What if we could communicate with the trees and the bacteria in their own language? The potential implications are far reaching.

MEDICINE

The oldest and the newest applications of Nature's technology are in medicine. Dogs licking the wounds of their human masters ten thousand years ago cleaned the wounds with their saliva, which contains antibiotic properties. Penicillin is Nature's technology. And pharmaceutical companies have been looking at mold and digging in the dirt ever since penicillin's discovery, looking for other natural compounds that they could improve upon.

Magainins. In the late 1980s, Dr. Michael Zasloff pioneered the use of natural peptides to kill pathogenic bacteria. He found his first antimicrobial peptide on the skin of an African clawed frog. He named this class of antibiotics Magainins (which is Hebrew for shield). He and other researchers have found similar peptides in sharks, moths, pigs, and jellyfish. Zasloff believes that this new category of antibiotics is revolutionary because pathogens have never been able to build resistance to it.

Maggots. Not only are Nature Technologists enlisting the secretions of living creatures, they are also enlisting the creatures themselves. Blowfly maggots eat only dead flesh. So doctors are using them to clean out bedsores, tumor-killed tissue, and certain kinds of burns in people who would be at risk in surgery. They have also been used to fight bone infection where antibiotics have a poor record of success. Maggot therapy is successful 90 percent of the time.

While this application of maggots is new to modern medicine, it is known that the Maya used them therapeutically a thousand years ago.

Leeches. The University of Michigan at Ann Arbor has used leeches to drain off old blood from transplant operations. It was successful in all fifteen cases where they were used, even though the transplants had been labeled "beyond saving."

Worms. A parasitic worm that causes schistosomiasis also prevents fibrous deposits from building up in arteries by breaking down fibrinogen, a protein involved in arterial blockages. Researcher Ronald Stanley of the University of Wales has discovered that the worm's presence also lowers cholesterol levels in its human hosts. He thinks that this benefit may explain why the body tolerates the worm long after the acute infection has subsided. The work of Nature Techers is to figure out how to utilize all the benefits of this worm without having to have the acute infection first.

Animal Self-Medication. Perhaps one of the most amazing phenonema of Nature Tech is the discovery that animals use parts of plants to medicate themselves. Chimpanzees have been observed picking a particular leaf from a particular plant, eating it, and several days later, defecating dead intestinal worms. Scientists checking the active ingredients of the leaves found they were toxic to those worms. A woman biologist watched a pregnant elephant walk dozens of miles beyond her normal territory, tear a small tree apart, and eat the bark. The following day the elephant went into labor and gave birth. When the scientist asked local villagers if they used the bark of that tree, they told her they used it to induce labor. These are just two of many documented examples of animals "consciously" using natural medicines to help themselves.

Human Growth Hormone (HGH). Human growth hormone is used to treat thousands of children around the world with growth problems. Recently Nature Techers have been using special bacteria to produce this hormone, but it is extremely expensive.

Dr. Daniel Salamone at the University of Buenos Aires in Argentina had a better idea. He took some cow cells growing in a Petri dish. He added to these cells human genetic material to produce human growth hormone. Then he cloned one of the cells and grew a cow. This cow, named Pampa Mansa, produces milk with human growth hormone in it. Ten percent of the milk's protein is HGH.

Now, here is the stunning result: Just fifteen cows like this one can produce the entire human growth hormone demand worldwide! This is what is happening in the Nature TechnEcology. And it is going to revolutionize industry after industry.

TRANSPORTATION

Right now, we can only look to the old Nature Tech solutions: horses, mules, elephants, oxen. But, Nature Tech is beginning to influence the shape of airplane wings. It turns out that the bumps on the fins of the humpbacks allow these big whales to turn sharply when chasing nimbler prey. Research led by Laurens E. Howle of

Duke University shows that the flipper with bumps is more aerodynamic than a smooth fin. Frank Fish, the scientist who first noticed this configuration, has patented the concept of lumpy lifting surfaces and believes it could have dramatic impact not just on wings but on propellers, helicopter rotors, and ship rudders.

COMPUTERS

DNA Computation. Perhaps the most dramatic application of Nature Technology is in "growing computers." It all started when Leonard Adleman, a mathematician at the University of Southern California, turned a beaker of DNA into a parallel processing computer. Even though the biological process is much slower than the computation speed of electronic computers, Adleman was able to enlist billions of strands of DNA to work on one problem simultaneously. The result: The DNA computer solved the problem in a matter of hours. If the most powerful electronic computer had been asked to solve the same problem, it would have taken months! This new way to compute biologically is a paradigm shift and could revolutionize everything in the entire computer industry.

DNA Tiles. Since Adleman's groundbreaking experiment in 1995, the field of DNA computation has moved steadily forward. In 1998 Nadrian Seeman and his colleagues at New York University figured out how to make tiles of DNA that assemble themselves into two-dimensional patterns.

Erick Winfree, a scientist at Caltech, has extended and improved Seeman's idea and has figured out how to program the tiles. These tiles are the forerunners of intricate nanostructures for biological computing and even biological manufacturing of nanotech structures.

From these examples, we can see a clear pattern of growing technological sophistication. Will Nature Tech someday "grow" television sets and lawn chairs? Since Nature has already grown tigers and red-woods and coral reefs, we think the question is not "Is it possible?" but "How soon?"

403 ▰
A Visit to the Nature TechnEcology

THE WORLD OF NATURE TECH IN 2050
What follows is a Web site summary of the biotechnological revolution that occurred during the last sixty years.

The paradoxes of the world of 2050 are abundant. At the midpoint of the twenty-first century, all of the predictions of doom have not happened. Yet the solutions to the impending catastrophes of pollution, resource depletion, global warming, and overpopulation did not come from human design and fabrication. Instead, they came from Mother Nature herself.

We learned to listen to her and use her ideas to help, not just us, but all life. We are finally at peace and in partnership with Gaia—our living planet—and we got here by declaring a truce.

Human beings finally figured out that Mother Nature is not the enemy. Although Mother Nature had always been trying to teach us the correct way to interact with her, we fought an ongoing war with her for more than ten thousand years that began with the Agricultural Revolution.

During the past fifty years, humans finally gave up on the Agricultural Revolution and its perpetual damage to the living systems of the planet. We are well along the way to building a long-term partnership using a new kind of technology, derived mostly from the natural processes that Mother Nature perfected over the past three billion years that allow all living things on the planet to flourish in their proper proportion.

This partnership began sixty years ago with very simple discoveries and has progressed through three stages. Each stage is distinct in the way humans and Nature interacted.

At the beginning of the twenty-first century, at least 99 percent of the world's microbes had never been cataloged, cultured, or examined for their benefits. So, with the advent of the new biochip testing equipment, which allowed ten thousand tests to be run simultaneously, it was like walking into a candy store. Almost daily, new discoveries were made.

All of these discoveries were important because they pointed toward an approach that had been ignored for the previous 150 years. Could it be that by simply looking at how Mother Nature had solved significant problems in the development of the 106 ecosystems around the world, humans could emulate, modify, and adapt those solutions to solve our own problems?

The answer soon became apparent. And since those early days in the twenty-first century, the more we have looked, the more wonderful natural solutions we have found to seemingly intractable problems.

We are now in the third stage of the development of Nature Technology. Each stage is summarized below.

STAGE 1: BIOMIMICRY (1990–2015)

Revolutions always start with a few key events. The general approach for the first part of the Nature Tech Revolution was biomimicry triggered by a book of that name by Janine M. Benyus, the Madame Curie of Nature Tech. Her recommendation: Find processes and products in nature and then mimic them with our manufacturing technologies.

Biomimicry made its first big impact with the discovery and cultivation of hydrogen-producing bacteria in 2008. Pollution from hydrocarbon fuels was becoming a planetary crisis, because of the CO_2-induced global warming. A multinational effort to increase the efficiency of hydrogen-producing bacteria led to the development of a class of engineered algae that produced large amounts of hydrogen from water. With this discovery, the world began to move toward a fuel source that created almost no pollution.

Medicines were the second most important part of the revolution. With the development of biochip testing equipment, the drug industry began to economically explore the medicinal qualities of literally hundreds of thousands of single-celled organisms.

And, while the drug industry focused on the biochip process for finding solutions, a group of environmental scientists went into the wilderness searching for "interesting" compounds. Magainins, an antimicrobial peptide discovered by Dr. Michael Zasloff on the skin of the African clawed frog in the late 1980s, became the paradigm of wilderness medicine and the first choice for treating a wide range of previously untreatable biological ailments.

More discoveries followed using his approach. In Australia, a set of complex chemicals found on the slime coating of sea slugs turned out to have an amazing ultraviolet-light–blocking capacity. Australian scientists quickly turned the discovery into a commercial sunblock, which has been credited with the dramatic reduction in skin cancers, especially the deadly melanoma.

From observations made by other scientists in the wild, it was clear that the animals were medicating themselves by eating certain leaves, berries, stems, and even soils. Brought back and tested, these animal "cures" turned out to have powerful natural compounds that worked against fungal infections, parasites, bacterial infection, and viruses.

While natural chemicals were being copied and adapted for medical use, hospitals began using insects, such as maggots from the blowfly, to clean dead tissue from wounds and burns. It didn't take patients very long to get over their initial squeamishness once they realized that the maggots were much more efficient and much less painful than any other choice available.

In the American heartland, farmers began to buy wasps to pro-

tect their corn in storage bins from attack by weevils and other creatures. The result was a dramatic reduction in damaged corn, which,
in turn, lowered the cost of food even as it improved the profits for
the farmers.

In the kitchens around the world, a new container began to appear underneath the sink. Organic waste was dropped into it and a
colony of three-inch–long red earthworms turned the waste into
odorless castings (more accurately described as worm poop) so rich
that gardeners refer to it as "brown gold." The city of Vancouver,
Canada, started the trend.

Plants—in particular, sunflowers—were used to trap soil pollutants in their roots so that, by harvesting the sunflower root, we could
also "harvest" the pollution and return the soil to its pristine state.
Bacteria were also enlisted for this work with especially difficult compounds such as chlorinated benzene chemicals. We discovered that,
for certain microbes, these toxic chemicals were food. The bacteria
"ate" them and converted them into harmless waste products.

Organic preservatives made from extracts of dried plums were
applied to fresh meats to kill *E. coli, Listeria,* and other nasty
pathogens. And certain vegetables and fruits became a legitimate
part of cancer-prevention programs.

Mollusks, under gentle investigation, gave up their formula for
waterproof glue. Geckos showed us how to utilize nanoscale whis-
kers to produce a dry tape that had enormous sticking power, yet,
when peeled off, left no residue.

While it was the chemical side of Nature Tech that got the most
attention in the early stages, there were indications of so much
more waiting to be understood and applied. One example was the
discovery and mimicking of the Stenorcara beetle's wing pattern,
which it uses for precipitating water out of the atmosphere in
Africa's Namib Desert, one of the driest places on Earth. Thanks to
this extraordinarily efficient design, desert villages that had previously
struggled to find enough safe drinking water could collect all they

needed from the morning air by using twenty-meter–tall versions of the beetle wing design.

A second example of design mimicry was the "bat wing" for sailboats. Richard Dryden, an English biologist, invented a sail for boats that opens and folds, altering its shape and size like a bat's wing to adapt spontaneously to new conditions with no need to change sails. It was clearly the precursor to our midcentury wing-ships. A third example of biomimicry was a new kind of microphone for hearing aids based on the "ear" of the *Ormia ochracea* fly. Its unique design gave engineers the template for building a tiny microphone that could help a listener identify from which direction the sound was coming. It also showed the engineers how to better filter out background noise, always a bane for hearing aid users.

But even as we were beginning to adapt natural technology using late-twentieth–century mechanical and chemical technology, we were also beginning to discover the wealth of solutions that were already available without such expensive adaptation. Colored cotton began to show up in the cotton fields around the world, the result of simple cross-breeding. There were benefits far beyond just the color. If your cotton is colored before you harvest it, then the entire process of dying is unnecessary. This, in turn, reduces the need for huge amounts of chemicals, the cost of handling the pollution those chemicals create, and the water needed to deal with the chemical dying process. The natural cotton-growing regions of the world became economically healthy because of this new approach to cotton fibers.

Biomonitors based on living organisms with natural sensitivities to various pollutants began to find their way into the marketplace. Mineral waste fields began to be mined by microscopic organisms that concentrated diffuse but valuable minerals into quantities that could be harvested simply by flushing out the waste fields after the bio-concentration was finished.

Biofilms, the scourge of humans throughout history, were enlisted to help solve some of the most pressing problems of the twenty-first century. Their first application was for the military. The return of biological weapons required a defense that was effective yet inexpensive. Biofilms turned out to be the answer. Biotechnologists were able to grow a living film from a combination of bacteria that defended the "wearer" from a broad range of pathogens. It even changed color to indicate which bioweapon it was fighting, and would signal a pulsating alarm of color if it was failing in that fight. Within five years, the countries that had developed the bioweapons signed an accord not to continue. It was clear that Mother Nature's defensive systems were better than her offensive systems.

Once biofilms became domesticated, a new industry was born. Biofilms were developed to shield people from sunburn. Biofilms were developed to protect metal and wood. Biofilms coated windows and controlled the amount of light coming through while keeping them clean. Biofilms became cleaners of clothes, floors, and transportation devices.

Insects were recruited for the revolution, too. For instance, honeybees were used to carry natural fungicides to the plants they visited, thus protecting the plants from disease. Honeybees also turned out to be excellent biomonitors of pollutants.

All of these natural innovations initially had only marginal effects on the technological superstructure of world culture. But even though these activities affected less than 2 percent of the global economic activity, they were doing something essential for a revolution to occur: They made people increasingly comfortable with this new kind of "Nature Technology," thus preparing them to be open to much more profound biological surprises that began to surface in the second stage of the Nature Tech Revolution.

STAGE 2: PRODUCTION PONDS (2015–35)

Plastic from plants—that surprised many people. Plastic had always been considered the most artificial of materials. But by 2015, the chemical giants of the world had figured out how to produce polymers from plant material. These "plant plastics" required 20 percent to 50 percent less energy to produce and composted into harmless carbon dioxide and dirt, thus solving one of the biggest problems of modern waste disposal. While the Cargills and the Dows were producing natural plastic chemically, other companies were teaching the plants themselves how to produce a wide variety of polyesters. By 2020 the majority of plastics in the world were "natural."

The next area of the revolution was in the manufacturing of ingredients that before had required huge expenditures and enormously complex structures. By the end of the '20s, many of these expensive plants were replaced with "production ponds." These ponds were filled with the necessary nutrients, kept at an optimum temperature with just the right amount of light from the proper part of the spectrum. Then combinations of microorganisms were introduced into the pond to "manufacture" specific materials. For example, cotton fiber began to be pond-produced by inserting the fiber-making part of the cotton gene into a single-celled microbe. With further adjustments, the strength of the fiber, the length of the fiber, as well as the color could be controlled and produced by the microbes on demand. The function of a $2 billion factory was captured in a microbe. And hundreds of billions of those little living factories operated in a pond.

The same process was applied to spider fiber. In the early 2000s, a company had tried to do this with goat's milk, but too many people felt that this was "udderly" unacceptable because of the use of a mammal to produce an insect's fiber. With the advent of the production ponds, this amazingly strong and stretchy material became a commodity aptly named spider silk.

A broad range of plastics, all biodegradable, was grown in the ponds. Even carbon fibers were produced in the ponds, thus making one of the strongest materials in the world cheaply available to anyone.

As soon as carbon fibers became commodities, they were integrated into almost all composite materials to provide high strength and low weight. Airplanes and trains, high-rise buildings and bridges, bicycles and furniture—all benefited from cheap carbon fiber.

Drugs of all sorts went through the same "pondification" process with the same result—lowered costs, faster production, and much wider access across the world. The remainder of the classic manufacturing process still had to be done: turning the medicinal material into pills and packaging them. But that was easy enough to do. The basics of healthy nutrition also became easily producible with the help of modified single-celled creatures and other more complex aquatic plants such as duckweed. Nature Techers experimented with thousands of combinations to optimize the process.

Other important commodities like soaps, cleansers, paints, adhesives, rubber, thin films, waxes, and paper fiber were produced in ponds.

By the end of the second stage of the revolution, our partnership with Nature had become so exact that we began to grow not just carbon fiber, but carbon fiber structures: parts for vehicles; beams for constructing homes, schools, and commercial buildings; legs for furniture.

Using the pond process, panels of very efficient insulation material were grown in exact shapes for easy assembly. Optical fibers were grown within the insulation panels for communications links.

In "microponds," the capacity to grow human organs also began to develop. Teeth, using the patient's own DNA, were grown to fix any dental problems. Skin, muscle tissue, and cartilage—all derived from the patient's own genetic code—were grown as needed.

All of these Nature Tech solutions had one thing in common: They were logical extensions of what was already being done naturally. It was in the field of computation that the truly hard-to-imagine occurred.

Near the end of the second stage of the revolution, DNA computers became the computational engines of choice for doing massive parallel processing to solve some of the most difficult problems of the era. Problems that would have taken years to solve with the electronic computers of the early twenty-first century were solved in less than twenty-four hours with DNA computers. By 2030 biological computing using DNA and its variations had become the dominant form of computing.

The natural computers were triggered by development of DNA tiles, which were conceived in the early 2000s and perfected by 2015. DNA tiling technology also was turned into nanotech assemblers. The initial research was started at New York University by Nadrian Seeman in the late 1990s. Leonard Adleman of the University of Southern California expanded Seeman's work and they formed a private company, AdleSee, Inc., which became the Microsoft of DNA assemblers.

With this technology, complete assembly of microstructures used in many of the communications devices of the day became easy to do. So, the reconceptualization of computers marked the increasing maturity of the revolution.

More technological substitutions continued. Bioluminescence, light generated by biological agents, moved from fireflies flickering in the summer night, to biological "light bulbs" that produced a soft glow. And the color of the glow was controllable depending on which variation of organic elements you had inside the bulb.

The capability of certain plants to be exothermic—to produce heat—also began to be harnessed and improved. Starting with *Philodendron selloum,* which in its native state can generate as much heat as a three-kilogram cat, Nature Techers ramped up its heat-producing capacity until clusters of them were able to be used as a significant heating source in well-insulated homes. Eventually, heating panels were grown using the DNA of exothermic plants as a template. Illumination panels, water purification panels, biological toilets, and waste-lets, organic devises for converting household wastes into soil nutrients, appeared as homes became more and more organic.

By the end of the revolution's second stage, the heating panels had added sensory elements to measure the room and the organisms in the room to optimize the temperature. The illumination panels also gained sensors. Water purification panels not only kept the water pure but gained communications capability to report unusual chemicals in the water. The biological toilets took measures of the urine and fecal material and sent data to the house computer to be shared with the medical support system as needed. Humans began dwelling in living structures that warmed us, cooled us, illuminated our work, helped us communicate, and fed us.

The revolution spread even to transportation. Nature Tech fundamentally redefined the transportation sector, once the engineers figured out the proper mode. It was dirigibles.

Dirigibles came back into full-time use except this time they were "biodirigibles." The skeletal structure was grown from carbon fiber. The organic surface covering fabricated itself over the lightweight carbon skeleton that had folding points throughout the design.

The passenger compartment was grown below the inflated structure in a beautiful aerodynamic form with organic windows whose DNA was borrowed from human corneas.

The hydrogen for the dirigible was generated by the organic dirigible fabric that converted solar energy into electricity and, combined with enzymatic catalysts, separated water into hydrogen and oxygen. Because the spars of the dirigible had joints grown into them, the dirigible, once it was tethered, could absorb the hydrogen gas into an aerogel inside the dirigible and fold to one tenth of its inflated size in a matter of minutes, making it much smaller and therefore much easier to store. When it was ready to depart, the hydrogen was released by heating the aerogel, expanding the envelope back to its full size.

The cost of a Nature Tech dirigible was a fraction of that of the airplanes they replaced. In many ways, it looked like a giant living thing. But, during the second stage of the revolution, intelligence was not yet built in.

By the end of the second stage, more than 70 percent of all products were produced using Nature Tech. Across the world, the production ponds changed the access to previously hard-to-manufacture products and materials. For instance, many parts of India skipped the old paradigm of industrialization completely and went right to Nature Tech for necessities of postindustrial civilization. China, on the other hand, took longer to reap the Nature Tech benefits because it had so deeply invested in mechanical and electronic industrialization.

Once the transformation was made, cost of living plummeted and the quality of life soared. The natural environment, which had been devastated by the Industrial Revolution, began its healing process.

STAGE 3: THE INTEGRATION OF INTELLIGENCE (2035)

It was the search for intelligent housing that drove the third stage of the revolution. As housing panels became increasingly clever, it seemed logical to give them greater "intelligence." By building in neural networks with the capacity to store and analyze the data, houses became smart in ways undreamed of in the twentieth century.

The dirigibles were next to gain intelligence. By adding the right kind of "nerve" tissue, these magnificent creatures of the sky began to develop sentient capabilities. Now the ship advised the pilot, kept track of the weather patterns, monitored thunderstorms, and anticipated times to turn and to climb and to descend. Our "partnership" with Nature was taking on whole new attributes.

Roads for ground transportation also changed radically. The new roads were grown organically and were rolled out in a matter of days.

They could heal their own potholes and cracks. And, with their new intelligence, they could weigh and count vehicles, track traffic patterns, and communicate with the necessary systems to make sure the roads ran in optimal conditions.

Then, in 2040, the integration of intelligence began to manifest itself in the new cities. In the Amazon forests, the Brazilian Nature Techers began to reap the benefits of their research to get trees to grow into homes. These arboreal cities allowed the forest to flourish even as it made natural space for humans to become part of the system. The tree cities reminded people of the twentieth-century drawings of the great city of Galadriel in *Lord of the Rings*.

Along the tropical seacoasts, coral organisms partnered with humans to construct incredibly intricate and beautiful ocean cities. Curved like giant shells and colored with organic materials, these cities glowed gently in the night from their own bioluminescence. These coral cities also created immense sanctuaries for the creatures of the sea, thus improving the abundance and diversity of ocean life. On the oceans, living wing-ships sailed. Their design, adapted from bat wings, allowed them to move with great efficiency. They were grown in the same way that the dirigibles were grown. And they were as smart as the dirigibles.

While Nature Techers had also done some space exploration using old technology, it was during the third stage of the revolution that outer space became significant. The first living space station grew into being during the ten years between 2030 and 2040. It spawned hundreds of other living structures that orbited both the Earth and the Moon.

Using Nature Tech, interplanetary sail-ships began to ply the vastness of space using thousand–square-kilometer sails to catch the sun's photons for motive power. The sails also converted some of the photons into electricity to power the ship's equipment. Because the ships were alive, any punctures in the sail were not repaired, but healed. The passenger compartments rotated to provide a sense of gravity and, of course, the ship grew all the foodstuffs necessary for the trips while purifying the air biologically. For deep space travel, a linear accelerator bioship was developed. Its design was basically a straight alimentary canal with organic magnets that could be pulsed. Asteroids were mined for iron, which was converted into pellets that were magnetically accelerated out the tube, providing Newtonian thrust for the ship. Very high speeds could be attained within weeks of departure.

On the Moon, cities were grown using the most sophisticated partnerships. The cities were created by tunneling under the surface. Then the protective layers were grown with lead material incorporated to protect both the living structure and its inhabitants from the intense radiation of solar flares. The great windows that looked out upon the lunar landscape had "eyelids" that could close in a flicker to protect from meteorite damage. Even the early-warning radar systems were organic and connected to organic computers.

As part of the partnership, humans finally understood and accepted their most important role: the protection of the home planet. With our natural partners, we have formed a ring of protection around the planet to ensure that it will never be struck by asteroids again.

The Nature TechnEcology is in full bloom at the halfway mark of the twenty-first century. The planet is returning to its full health. Humans have found their future in a partnership with all the living things of the planet.

5

HUMAN TECH

501
Overview and Guidelines

What lies behind us and what lies before us
are tiny matters compared to what lies within us.

RALPH WALDO EMERSON

We are now at the last of the five regions. This region, Human Tech, is so unlike the rest of the regions that you might, at first, consider it an error in cataloging. The first four regions are constructed of materials and, with the exception of medical devices or products, operate outside the skin of human beings. Human Technology, on the other hand, operates, for the most part, inside our skin. We find it in both our physical capacities and our mental capacities. Our understanding of both of these domains has grown exponentially in the past twenty-five years.

The more obvious of these domains consist of the physiological tools that we have been endowed with through our genes. Only recently have we begun to understand the scope and power of these

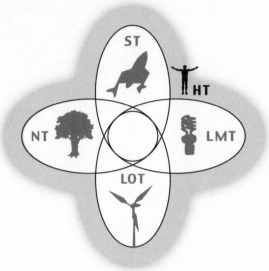

technologies in any conscious way. The second domain of Human Tech is made up of those metaphysical tools created by our minds that have allowed us to organize our species, discover the way the world works, and create new ideas in art, culture, religion, science, and technology.

So the fifth region is located in a totally different place, not in the external world but in the internal world. But, that internal world is still a TechnEcology with separate technologies working interactively to create a complex system for solving problems—the problems of being human.

That makes the fifth region the most important, because the other four regions are manifestations of the need to solve problems for the Human TechnEcology. Because of its basic difference, we have to represent this region in a new way. As you can see in the Venn diagram above, we have chosen to depict this region, not as a single oval, but instead as a region in which the others reside. This region enriches, encompasses, and enables the other four regions. In a metaphorical sense, the Human TechnEcology is the ocean from which the other four, like interlinked islands, arise.

Human Technology is a very new and very old technology. As with Nature Tech, humans have been using some of the technology in conscious ways for tens of thousands of years. But during that time, most of our Human Technologies have operated beneath our consciousness. Only in the past three decades has mainstream science focused its gaze on the nature of these technologies. Because of that research, we have moved from superstitions, old wives' tales, and the shamanistic "magic" to a much clearer understanding of human capabilities.

So, the technology we are examining in this region starts at the marrow of our bones and the center of our brain, and extends to the ends of our imagination.

Because Human Tech is the result of a set of tools that we are born with, no one of us gets every tool or even the same tools with the same sharpness. The edges of our tools differ in their cutting power depending upon our genetic heritage and our educational and cultural upbringing, which brings forth and focuses our genetic attributes.

Guidelines

Here are the four rules of the Human Tech region:

1. *The real needs of humans are* not *material needs.* We don't *need* an automobile or a straw bale house or an organically grown wardrobe. Which material options we choose isn't the issue. Our true needs are linked to internal states—our hopes, our fears, our dreams, our relationships, and our goals.

2. *Science is only now learning how to measure human technology.* With the recent advances of the computer and a growing array of sophisticated monitoring devices (like magnetic resonance imaging), we have developed the tools that can sense and measure our hidden capacities. Psychology and sociology

have developed experimental methods that have begun to uncover the way we think, act, and interact, both individually and collectively. And other sciences, especially evolutionary biology, have begun to contribute to our understanding as well.

3. *God or Mother Nature or evolution, depending on your point of view, has endowed us with extraordinary capabilities.* This guideline puts us into a special relationship with our creator/evolver. The problem for the past several hundred years has been that we thought of ourselves in simplistic ways. We were *tabula rasa*, blank slates according to seventeenth-century philosopher John Locke. For behaviorist B.F. Skinner, we were nothing more than an accumulation of conditioned responses. These simplistic points of view have been overwhelmed in the past twenty years with research that suggests our physiological and psychological capacities are much greater and more sophisticated than we ever imagined.

4. *Our true work is to know ourselves.* Socrates had it right over two thousand years ago. Until we understand our capacities, we don't know our true potential. In the Human Tech region, the most important work is to discover, describe, and understand the human being in all our glory.

The Human Tech slogan is:
"We are beautiful!"

502
Advocates and Examples

And the narrowest hinge in my hand
puts to scorn all machinery.

WALT WHITMAN

While the region of Human Technology has been an ongoing discovery-and-development effort from prehistory, only in the past century has it been rigorously explored by science. Small forays by pioneering scientists began in the 1920s and '30s. But Nazi human experimentation in the 1940s put human research on hold for almost two decades. It wasn't until the late 1950s, led by biofeedback researchers, that serious exploration began again. Since then, the range and the depth of that exploration have grown steadily. Researchers from many fields have made major contributions to the understanding of the human capacity, both physical and mental. We have chosen a representative sample from both areas to illustrate just how large this region is.

Advocates

Elmer Green. The author of the classic *Beyond Biofeedback*, Dr. Green, through his work at the Menninger Institute in Kansas City, broke open the floodgates of research on biofeedback. His quantitative research with a master yogi from India proved that a human being could actually "control" certain autonomic processes of the body, which the yogi demonstrated by stopping and restarting his heart with his own thoughts.

Thomas Bourchard. A researcher at the University of Minnesota, Bourchard did the seminal research on twins separated at birth. His work demonstrated that many supposedly cultural traits of human beings are actually influenced dramatically by genetic makeup. This work opened new windows on how important our inherited characteristics are in determining what we become.

Franz Halberg. Another groundbreaking researcher at the University of Minnesota, Halberg investigated chronobiology and revealed how important internal body clocks are in influencing human beings' performance. His work showed, for instance, that the effectiveness of medicines is dramatically influenced by circadian rhythms (the twenty-four–hour cycle). His work also discovered a wide array of rhythms from daily to multimonth, each of which influences human performance.

Randy Thornhill. A researcher at the University of New Mexico, Thornhill was the first to see and measure the connection between human symmetry and attractiveness to the opposite sex. This connection is explained by the fact that high bilateral symmetry is an external indicator of high-quality genetic health. Thornhill's studies indicate that this ability to sense symmetry, since it can be found in all cultures, is hardwired into humans.

James Thomson. A professor at the University of Wisconsin Medical School, Thomson was the first to isolate and culture a human embryonic stem cell line. Thomson's discovery and related research by John Gearhart at Johns Hopkins and Geron Corporation has spurred a great deal of additional scientific as well as commercial activity. Ethical concerns have been raised about how to proceed with the science, but there appears to be little doubt that stem cells are a Human Technology with great potential.

Martin Seligman. The author of *Learned Optimism,* Seligman has conducted research showing how profoundly attitude impacts all other aspects of human behavior. He has identified the "technology of hope." Again, if we think about technology as tools to help us solve problems, Seligman has identified key tools to encourage hope. His research shows that being optimistic gives a person substantial advantages over a pessimist. Perhaps the most important message in Seligman's book is that one can learn to be optimistic and gain those advantages that come with a positive attitude.

W. Edwards Deming. The father of the Total Quality (TQ) movement, Deming—along with Joseph Juran and Phillip Crosby—created a technology of procedures and processes that enables people to produce almost error-free products and services. This Total Quality "technology" is an example of nonphysical technology, technology of the mind. As a tool, TQ allows humans to work more effectively in the first four regions.

Peter Drucker. His greatest contribution in Human Technology is his theory of management. The author of too many books to list, he organized and structured the process for the effective management of organizations of all sizes. His systemization of management technology changed forever the way organizations are run. Check the bibliography for more Human Technologists.

Examples

PHYSICAL

Bilateral Symmetry. How do women pick good mates? The old paradigm focused on obvious measures such as size and strength of the potential husband and his assets. But Randy Thornhill, a biologist at the University of New Mexico, and Steven Gangestad, a professor of psychology at the University of Vienna, have found that women have a more powerful measuring tool.

In their research they asked women to choose from a set of pictures a man they would want to father their children. When women were given their choice of handsome but asymmetrical men and less handsome but very symmetrical men, they almost always went with the symmetrical man. Women made the same choice regardless of their culture.

Why would symmetry be so important that it is hardwired into the human brain? Other researchers have demonstrated that bilateral symmetry is the clearest external indicator of genetic health, not just for humans, but for all creatures that are bilaterally symmetrical. So, when a woman picks a highly symmetrical man, she is picking the best genes available, which, in turn, increases the likelihood of healthy offspring. This is an example of the internal measurement tools that are part of our Human Technology.

Skin Color. What is the connection of skin color to Human Technology? Research reported in *New Scientist* provides an answer. The old assumption of skin pigment was that it protected us against sun damage. Researcher Nina Jablonski from the California Academy of Sciences suggests another, more important reason: "Skin color is not the result of a simple one-to-one relationship between ultraviolet light and melanin."

It turns out that skin color has two functions. Being dark has benefits because it protects the skin from UV damage and prevents the destruction of folate, one of the members of the vitamin B complex. Folate is essential for human health, so protecting it from UV breakdown is critical. But paleness allows the sun to penetrate the skin and helps us make vitamin D_3, which is just as important for human health as folate.

Skin color is a balance between the protective needs and the production needs of the human system. And, as is so typical of all Human Technology, skin color is much subtler and more sophisticated than we have previously thought.

CHEMICAL

For most of the twentieth century, medicine used chemical analysis of urine and fecal material to measure a person's health. Now we are finding that other substances our bodies create are sources of information about our health.

Earwax. According to epidemiologist Nicholas Petrakis at the University of California at San Francisco, earwax contains evidence that may indicate a woman's risk for contracting breast cancer.

Tears. Tears, too, hold important chemical information. They also may carry away emotional toxins that accumulate as part of the grief process. It is important to note that to measure the information in these secretions, we need very sophisticated technology. But, the information itself is created within the body and released through these channels. All that has been lacking was the proper measurement equipment.

Pheromones. Pheromones, odorants produced to signal specific kinds of information, are produced by humans. Research done at the University of Texas, Austin, found that men preferred odors from women who were ovulating to those same women's odors when they weren't ovulating. In other words, men could differentiate, by smell, when a woman was ready to conceive—and that smell was more desirable. This kind of chemical communication is just one of many that are being identified scientifically. In all likelihood, we have just begun to uncover a lexicon of pheromones that humans use to influence one another.

Mother's Milk. The Super Techers thought their substitutes, Infamil and Simulac, were just as good as the real thing. The truth is that those chemicals are not even in the same league with Mama's product. For starters, breast milk comes equipped with antibodies that kill rotavirus, the most common cause of diarrhea in infants. It also contains complex carbohydrates that protect against other disease-causing organisms. These natural additives protect the baby from herpes virus and bacteria such as *Helicobacter pylori* (which is known for causing stomach ulcers).

And, in a demonstration of profound customization, mother's milk changes its chemical makeup daily when the mother is nursing a premature baby. We might want to ask ourselves how the mother receives the correct information from a baby to change the physiological processes of her body to make the personalized nutrients that particular baby requires. We don't know yet. But, for sure, it will be another example of Human Technology.

Gene Therapy. Perhaps one of the most important advances in medicine is coming from gene therapy: the use of human genetic material to cure disease or disabilities. For instance, to help with severe chest pain, patients were injected directly into their heart muscle with a gene that stimulated the growth of new blood vessels. The new vessels bypassed the obstructed arteries, increased blood flow to the heart, and dramatically improved heart function while lowering the pain.

This research adds to an emerging body of evidence that indicates gene therapy can trigger new growth or corrective growth by enlisting the body to fix itself.

Chronobiology. Timekeeping inside the human body requires internal clocks—clearly Human Technology. Circadian rhythms, for instance, are our twenty-four–hour clocks. Franz Halberg, at the University of Minnesota, has been studying biological rhythms for more than three decades. He has mapped the rise and fall of many indicators in the body. Hormones, blood pressure, body temperature, to name just three, rise and fall in regular cycles over twenty-four hours.

Why is this important? As an example, Dr. Halberg has discovered that medicines are most effective at certain times of the day. In the case of some cancer medicines, taken at the "wrong" time, it would require twice as much of the prescribed dose to be only half as effective. Yet taken at the "right" time, half the dose can be twice as effective. Given that many cancer medicines are very toxic, the timing can make an enormous difference in how much medicine has to be used for successful treatment.

Chronobiological research done by Jeanne Duffy at Boston's Brigham and Women's Hospitals reinforces what most of us know about ourselves: There are "early birds" and "night owls." Other researchers have found that we can solve problems better during certain times in our daily rhythm and that we absorb information better at specific times. Chronobiology is an emerging Human Technology still waiting to be applied intelligently.

Stem Cells. The discovery of stem cells, primitive body cells that can change into any specific cell in the human body, has begun to revolutionize medicine. This field even has its own name: regenerative medicine. Already, stem cell research has led to the ability to grow new corneas for people whose corneas are damaged or diseased. Researchers in both California and Taiwan not only have grown new corneal material but have transplanted the material into humans. The result has dramatically improved their vision.

Stem cells hear the call of damaged cells and migrate to the area where they can work their power of regeneration. Research done by Barbara Tate of Children's Hospital in Boston tracked stem cells as they found their way to damaged areas in the brain. And, in early 2002, University of Minnesota researcher Catherine Verfaillie announced that she had found stem cells in adults that can turn into every single tissue in the body. Because her claim is so extraordinary, it is being carefully checked. But, if true, it would mean that the field of regenerative medicine is poised to take off.

Antibiofilm Secretions. Biofilms are a major medical problem. They are highly resistant to antibiotics and can cause lethal infections. Research done at the University of Iowa has discovered a protein called lactoferrin in tears, mucus, sweat, and human milk that stops bacteria from being able to form biofilms. Lead researcher Pradeep Singh points out that on "every vulnerable human surface: lactoferrin is there." This discovery of another Human Technology may point the way to stopping pathogenic biofilms from forming.

METAPHYSICAL

What we have described so far in this section are examples of physical Human Technology. The next set of examples focuses on the mental components of Human Technology. Before we begin, let's revisit our definition of technology: It is a set of tools and techniques, processes, and procedures that can be used to solve problems. The range of examples is so vast in this area of Human Tech that all we can do is sample a few to indicate the scope of these technologies. The next two examples are transitional examples in which the mind creates a physiological effect instead of the physiology actually existing. Then from these two examples we move to the metaphysical examples of the mind creating new tools to enhance human performance.

Placebos. One of the most controversial indicators of Human Technology is the "placebo effect." Placebos have been used in medical research for almost one hundred years. The process is simple: Give actual medicine to one patient, give a placebo (a sugar pill, for instance, with no medicinal value) to another patient, and compare the effects. It is essential, however, that the patient taking the placebo believes it is the real medicine.

The assumption was that a sugar pill would not cause any change since it had no active ingredients. But statistically that wasn't the way it worked out. In fact, about 30 percent of the patients taking the placebo experienced measurable therapeutic benefits. How could this be? The most current explanation is that the patient's belief in the medicine triggers an internal healing response created by the patient's own "technology."

Recent research on Parkinson's disease showed that patients increased their production of dopamine, a brain chemical that reduces the Parkinson's shakes, after receiving a placebo. "The magnitude of the response is striking," said A. Jon Stoessl, who led the brain-imaging study that monitored the dopamine production. Exactly how the placebo technology works is still a mystery, but that it works is no longer in doubt.

Hypnosis. This topic is full of myth and confusion. But careful research indicates a set of capabilities in humans that are illuminated when people are hypnotized. If you are hypnotizable, you may be able to decrease your sense of pain drastically. In some cases the degree of pain relief achieved by hypnosis is equal to or greater than that achieved by morphine.

People who have been hypnotized can be told to hear things that don't exist. In brain studies using PET scanners, the brain pattern under hypnosis matches the pattern of the person actually hearing something. If the person is asked to imagine hearing the sound but is not hypnotized, the brain pattern is very different.

People under hypnosis can be told that ammonia has no smell, then sniff it, and not react to the fumes. This experiment demonstrates that human beings' thoughts can have profound physical influence. And yet, until recently, such a statement would have been considered fraudulent. It is another indication of how little we really know about ourselves.

Organizational Management. The father of this Human Technology is Peter Drucker. He systemized and explained a set of procedures and practices that allow organizations to better utilize their resources and assets and take advantage of opportunities in the environment. It is important to note that none of his management technology requires special physical equipment. It is all done with ideas. (That does not, however, suggest that computers should not be used to make Drucker's methods even better. But Drucker's methods do not require them.)

Micro Loans. Invented in Bangladesh, this Human Technology created a mechanism to effectively make very small loans to very poor people. Introduced throughout the country by the Grameen Bank, micro loan technology combines access to reasonably priced loans, education about how to run a small business, teams of borrowers who help one another, and the belief that poor people are as good a credit risk as the rich. Ninety-four percent of the borrowers are women.

Where they have been adopted, micro loans have fundamentally changed the lives of Third World villagers. A small loan, sometimes no larger than $12, allows someone to buy the necessary equipment and materials to dramatically improve his or her productivity and profitability. Perhaps the most telling statistic is that loan repayment runs between 96 and 100 percent—as good a figure as for the rich!

Teamwork. This is probably one of the oldest of Human Technologies. Every successful culture has developed some version of it. From the teamwork needed to bring down a mastodon to the teamwork needed to get a 787 ready to fly, the basics have never changed. Today, we realize how important teamwork is to the most modern factory or the most effective service organization. What we have been working on over the past decade is to optimize the process within the context of modern communications.

Leadership. Is leadership inherited or learned? Of course, the answer is both. Leaders have a certain smell to them (that's a result of specific pheromones), but they also do certain things that make others follow them. The number of books written on the topic recently is in the hundreds. It is still considered an art, but we are becoming clearer on the basic components of leadership and how people can be prepared for the role. The Center for Creative Leadership is one institution that has been developing a system of leadership preparation. And with a system comes technology—in this case, the technology to train a person to become a better leader. Obviously an entire book could be written on just this region.

503 ▮
A Visit to a Human TechnEcology—Schooling: 2030

We listen in on a planning meeting at a life preparation center in the year 2030. The principal is leading the meeting. His name is Frank.

"All right, learning leaders, please sit down. We'd like to get through this meeting, so we can go home early.

"As you all know, this exercise is done each year to remind us of what is happening with all eight age levels here at Cedar Creek Life Prep. Some of you are new this year. And some of you are forgetful. I consider myself in the second group."

Laughter comes from the sixty people in the room. They represent a mix of ages, genders, races, and experiences. In the group is a nurse, a child psychologist, a physical fitness and meditation expert, a nutritionist/chronobiologist, a linguist, a team management and leadership expert, and assorted specialists in reading, speaking, thought presentation, music and math (the two being considered the same subject these days), strategic and visionary thinking, foreign languages, parent coaching, friendship and bonding, ethics, and IQ/EQ/AQ (intelligence quotient, emotional quotient, adversity quotient). A special team formed from the group is responsible for measurements and standards.

"We asked you to prepare, as part of your update, a little history of your specialty, and then explain anything new that you will be doing this year. Let's begin with our health team."

Bob and Lisa stand up and begin their presentation. On the wall, in front, their key points and graphics are illustrated in AppleSoft Power Point 10.

"First, a little history on the reasons we run our days the way we do and how we illuminate the school," says Bob. "Believe it or not, it wasn't until 2005 that life prep centers—they were called schools back then—started looking at the circadian rhythms of their students. They were actually trying to start school as early as 7:30 A.M., long before their children's rhythms were ready for learning. As a result, they wasted the first two hours of schooling. Amazingly unproductive. And frustrating for the children who really needed their sleep.

"Besides that, most of those schools were illuminated with what was called fluorescent lighting. Today the spectrum of that kind of light would be considered hazardous to your health. This year, we will be adding some new controls to our full-spectrum LEDs to compensate for their age. They were installed in 2010 and should be good for another twenty years."

Lisa takes over: "As you know, all of our children have been mapped for their chronobiological rhythms so their learning events can be synchronized throughout the day. We have a new chronobiological measurement tool that we will be using to check several of the more subtle daily measures, and to improve our multimonth rhythm mapping. You all have your students' chronomaps in your database to help you structure testing and problem-solving times accordingly.

"We will also be collaborating with Mary, our new math and music teacher, to use the new data as part of the section on statistics. That has worked in the past and the kids have always enjoyed doing the math on their own data."

(Throughout the life prep curriculum, it is common to double up the work. For instance, children need to understand their own body time. The students use the data collected to help plan their education as the base to learn statistics, percentages, and standard deviation, as well as the chemistry and biology of the body. In the old paradigm of education, each of these topics would have been handled separately and teachers would have complained of not having enough time to teach. By using "double-up learning," that is never the case in 2030. By combining learning experiences, there is always enough time to learn everything!)

Lisa continues, "Now, we all know that children do best at problem solving and creative endeavors when they are at peak temperature. But this year we are going to expand the number of problem-solving experiences during their lower-ebb times, so they can experience their performance during suboptimal conditions. After all, there will be times in their lives when they can't schedule their problem-solving efforts for the best time of their day."

(As Lisa talks, one teacher whispers to another, "Isn't it unbelievable! Back in the twentieth century, the schools were actually structured almost exclusively for the morning peakers. The afternoon and evening peakers were never at their best when they were being tested or asked to do high-level problem solving throughout their entire schooling! It was incredible discrimination, but no one really appreciated the importance of chronobiology and its impact on learning until the 2008 lawsuit." Her fellow teacher nods his head in agreement.)

"So, in summary, we will be continuing the 9 A.M. school start and 5 P.M. finish, testing and problem solving at appropriate times, and introducing the new circadian measurement tool to fine-tune the student database. We also invite you to pick up a new wrist monitor and update your own chronomap."

Lisa turns back to Bob, who takes over.

"Now, what's new in nutrition? The children will be using new software to track their eating patterns. If you haven't looked at it, you'll find it under 'new software' at the site.

"We're ten years into the Optimizing Food System (OFS), and we continue to see dramatic reductions in obesity and other diet-related illnesses. Historically, you know, children were not well educated about nutrition because back then people felt that children were only interested in sweet and fatty foods. Of course, they were completely wrong! Children are fascinated with the diversity of food they eat.

"When you think about it, it is amazing what the nation used to spend in illness correction that could have all been avoided with proper nutrition and early prevention. For instance, all the refined starches and sugars caused systemic problems that manifested themselves in things like bad eyesight, obesity, and diabetes. People were always complaining about the costs of healthcare, but they weren't willing to make sure their children ate properly, which would have dramatically diminished health costs later on.

"One more thing: We will be adding some new varieties of fruits and vegetables to the center's garden this year for the children to plant and tend. As part of the OFS, we will do the nutritional analysis of these new foods to see where they fit into the system. Lots of good 'double-ups' for chemistry, biology, statistics, and anthropology in that exercise, content experts!

"And, of course," Bob continues, "the cooking teams will be assembled from your recommendations next week. This year, our children will be learning how to prepare new foods from twenty different countries. I've tasted some of the recipes already and they are fantastic! Bon appetit!"

Lisa stands up again. "Since our time is up, we'll stop. But please come by to see what else we're up to. We'll be serving tepenade and fresh sovlaki this afternoon at three in our area."

Frank takes over the meeting again. "Thank you, Lisa and Bob. Can't wait to taste what you've made. Next, we'll hear from the Visioning team."

Three people—Enrique, Mary Lou, and Hasim—stand up and move to the front of the room.

"A little history to start with," Hasim begins. "While corporations learned the value of positive visions of the future in the late twentieth century, it didn't become part of the learning curriculum until the twenty-first century. It was Martin Seligman's work on optimism that helped to convince schools that children with an optimistic vision of their own future learned quicker and better.

"We now know that a positive future vision not only affects learning, it also affects health by improving the immune system. Children who have learned to interact with the world in an optimistic way and who also have a powerful positive vision of their own future regularly outperform their measured capabilities. Their scores on the Competency Criteria reflect their vision commitment.

"We have received some new material for our Adversity Quotient training from the Stoltz Group. It looks especially useful for some of our youngest children. It will focus on building determination to work through hard situations. We will be using the climbing wall to create some of those situations.

"We are also responsible for the spiritual development of the children because vision and hope are intertwined. We will be having visits from a wide range of spiritual leaders, who will be talking about their experiences and the power of hope. These visits will expand our diversity learning for our children as well."

Mary Lou chimes in, "We will be adding more yoga work this year. Both Hasim and I went for further training during the last break. We have also added some new biofeedback stations. We invite you all to come and try them out. The training images are quite wonderful. And, during October, we'll be focusing on the bone and muscle charts. We've got a new holographic projection that not only highlights each bone and muscle, but also shows the range of movement and illustrates what happens when you damage that particular part.

"The meditation and prayer room got new rugs, thanks to the grandparents' group. And the new color on the walls is from the latest research findings. We think it definitely improves the quality of the space. That's about it."

"OK. Any questions for the Visioning team? Moving on, let's hear from our Relationships team," says the principal.

Christine and Charlie come to the front of the room. They're the rah-rah learning leaders of the entire center.

"OK, OK," Christine says as she clicks on their first Power Point. "History, huh? We've got lots of it. But just a couple of key points: For the longest time, old school educators thought that teams were for sports and that intellectual collaboration was cheating. Kids were dissuaded from working together by grading systems that were competitively based on statistical curves instead of on competency. Can you believe that? How did they ever expect for kids to learn to work together as adults when they measured them against one another as kids? Times have changed! As you know, we cover the following areas . . ."

She clicks on a new Power Point. It reads:

Teamwork
Leadership
Followership
Cooperation/collaboration
Mutualism
Total Quality
Productivity
Community work

"Since more than 70 percent of all work performed in society is done in teams, our centers are also structured the same way. This year we are going to be emphasizing physical teamwork through a series of Outward Bound experiences. We will be taking about one hundred children at a time to the Minnesota North Woods for a one-week teamwork session. The younger kids will be going in the spring and fall, and the older ones will go during the winter.

"This is the first time in five years that we have done a winter team experience. Of course, that means five of you will also be coming along for each of the sessions. Check the Web site for dates, and please sign up. Of course, if you don't get around to making your own choice, the old scheduler will get you!"

Charlie takes over next. "While those guys are up north freezing their noses, we will be doing some great new leadership and followership tasks back here. We have set up a series of interactions with the elders. We have several elders who went through the second Gulf War, the Terrorist Wars, and the Saving of the Oceans. They will talk about what they learned about leadership and followership in those experiences.

"We also have a new Web-center relationship in Curitiba, Brazil, so we will build our mutualism tasks around that relationship this year. If all goes well, we will be flying one hundred kids to Curitiba during the March break to visit.

"Our 3Ds program this year will focus on the Tor Dahl productivity process. If you recall, we used it two years ago to solve a series of curriculum problems. This year, we are going to focus on ways to increase the productivity of our students as they clean the building and grounds.

"We will also be continuing our Deming Quality efforts in our neighborhood painting program.

"And, of course, the Drucker volunteers' design program will be used by our students as they create the new vision of our center. They will be giving their recommendations in three months.

"All in all, we think it is going to be a great year. You are all part of a super team. And we think we model pretty well what we want our kids to behave like. We try to show them every day how the strength of difference is at the base of a democratic institution like our center. Let's give ourselves a hand!"

The room explodes in laughter and applause. The learning leaders know that they are among the most highly competent educators in all of history. And they know that because of the performance of their students. After all, that is the clearest measure.

"Good report. Thank you," says the principal. "Now we'll get an update and a little history from our medical staff. Nurse Thom?"

"Thanks, Frank. Good morning, everyone. You're looking good. I assume you are all up to date on vaccinations and health scans. If not, please come see me for an appointment.

"We have no new equipment this year. We really didn't need anything. A little history to think about: It took the United States of America almost one hundred years to figure out that the best time to form healthy attitudes and healthy citizens is in the first ten years of life. Schools had nurses, but were almost completely passive in how they dealt with children's health. They did NOT take care of their dental needs; they did NOT take care of their nutritional needs; they did NOT take care of their emotional needs.

"Instead they assumed that the parents, almost all of whom had had no formal parenting education, would do these things. As a result, damage that could have been corrected for relatively little expense was allowed to bloom into enormous social and personal health problems. The cost of correction was usually a hundred to a thousand times as much as the cost of prevention. Just another reason why it cost so much to make twentieth-century society minimally successful.

"So, because we have an integrated health approach that uses physical fitness and spiritual development and proper nutrition coupled with social involvement and positive visioning, and, because we start the health process so early, our national health is dramatically better now than thirty years ago.

"Today, in 2030, our society is five times healthier than in the year 2000 and yet it costs us 80 percent less to provide good health. The average American is healthier than the top 5 percent of healthiest Americans at the turn of the century.

"I am so proud of the fact that we take care of our children and we educate our parents in such a way as to generally improve both health and happiness. You see, Thomas Jefferson was right to speak of the pursuit of happiness in the Declaration of Independence.

"I think you all know the new additions to the National Health Care program, so I won't review those except to point out that as long as we all practice our Best Living protocols, we can expect to see national health costs continue to decline and national health improve.

"Boy, that sounded like a sales pitch instead of a report," Thom concludes and then sits down.

"World Connections is next," says Frank, the principal. "Who wants to present, please?"

The World Connections group is a combination of Web experts and content experts in foreign languages, history, world literature, and physical, political, and ecological geography.

Shawn steps up to the front. "We've been working on some really good stuff for this year. We will, of course, continue learning in the basic six languages—English, Spanish, Chinese, Russian, Arabic, and Global. We have built some new links to the best Web sites in these languages and will also have Web-pals so all the children can practice their languages. We are adding Navaho this year as another North American language. Last year the focus was on L'akota. And we are going to be learning some Portuguese this year, because our world geography focus is on South America. That will work well with our Curitiba connection in Brazil. And, just as an aside, great food in Brazil. Wonderful ingredients for the kids to work with!

"We have scheduled the Unified Learning Event for the last week in April, which will pull all the South American learnings together into one full school report and demonstration.

"Oh, I forgot some history. The history expert forgets to do history . . . right! OK, here's the one history lesson for today: The greatest nation in the world didn't think it was important to know very much about its closest neighbors until twenty-five years ago. Children were taught more about France than Mexico; more about Greece than Argentina; more about Russia than Central America.

Funny, isn't it? Not even wanting to know about your neighbors. Guess we fixed that, didn't we!"

"Thanks, Shawn. Let's take a short break and then plan to be finished by 2:45," says the principal. The group stands up and heads for the patio, where refreshing liquids and healthy snacks await them.

6. CONCLUSION

601 ■■■
The Universal Technologies

You may have noticed that certain technologies appeared in more than one region. In fact, some played a significant role in all five regions.

These are the Universal Technologies.

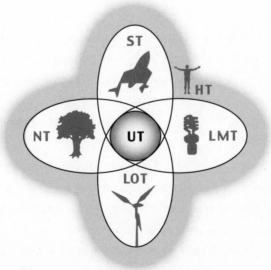

If you look at the five-regions diagram again, you can see that Universal Technologies must exist because of the overlapping nature of the regions. Some of the technologies overlap two regions, fewer overlap three regions, even fewer overlap four or five regions. When there are three or more overlaps, we consider them Universal Technologies.

A Universal Technology is the equivalent of the Swiss Army knife—so handy and flexible that almost everyone finds a *dominant* use for it. But it is also important to understand that the aggregate of the Universal Technologies does not equal another TechnEcology. Universal Techs do not cluster around any particular set of values.

For example, let's look at how each region might use the Internet:

The Super Tech region—to access a superabundance of information.

The Limits Tech region—to substitute for physical travel.

The Local Tech region—to exchange information and ideas between localities.

The Nature Tech region—for research and analysis of genomes.

The Human Tech region—for long-distance relationship building.

People in each region are enthusiastic users of the Web, but their basic motivation is different in each case.

It's the same World Wide Web, but its most important purpose is different from region to region. We don't mean to suggest that each region wouldn't also use the Web for other purposes, like access to vast amounts of information. But the *dominant purpose* is clear.

If you are a provider of a Universal Technology, your market is all of the regions. And because each of the regions uses your technology for a different purpose, you also gain a broad range of stimuli for further innovations of your technology.

Universal Technologies have the broadest base on which to grow and flourish. The key is understanding how to match your products to each region.

From our inventory, we have picked a dozen examples of Universal Technologies. With each one, we have described the dominant function they provide in each region.

Twelve Universal Tech Examples

1. Aerogel: This highly efficient insulation is made out of almost nothing.

> *Super Tech application:* It's used for wrapping space colonies to insulate them from extreme temperatures and to protect from micrometeorites without adding significant mass.
>
> *Limits Tech application:* It's used in all appropriate situations where insulation is required: refrigerators, homes, car roofs, water heaters, winter coats, and all sorts of industrial processes. An additional Limits advantage is that it requires very little material for optimum effect.
>
> *Local Tech application:* It's used for home insulation, in particular for insulation between glass panels in south-facing solar garden rooms. This allows plants to receive diffuse light through the aerogel windows while keeping the garden rooms protected from large swings in temperature during the winter and summer seasons. Aerogel can also be used to cover outside fish ponds at night with floating panels that can be easily lifted on and off because of low weight.
>
> *Nature Tech application:* Aerogels can be grown organically and used for the same applications as in Limits and Local Tech.
>
> *Human Tech application:* Nothing significant.

2. Thermal Depolymerization Process: This chemical process converts any carbon-based waste into gaseous and liquid fuels while capturing purified minerals that can be reused in manufacturing. It does this at 85 percent efficiency.

Super Tech application: It's used to convert junk and waste products of a superabundant society into more liquid fuel to propel personal transport vehicles.

Limits Tech application: It's used to turn most waste products into useful fuel (without adding more fossil CO_2 to the environment) and elementary materials. Even more important, it's used to mine old landfills and clean up the environment.

Local Tech application: It's used to convert local wastes into environmentally friendly fuels and materials to be utilized by the local community. This dramatically increases self-sufficiency while reducing dependence on distant energy suppliers.

Nature Tech application: It's used to convert organic wastes into fuels and materials (a transitional technology with only ten to fifteen years of usefulness). Emerging innovations in biological processes will render it obsolete.

Human Tech application: No direct use.

3. Computers: This includes single supercomputers, large computer arrays, and DNA computers.

Super Tech application: They can generate, manage, and make accessible to everyone the total information base created by all human endeavors.

Limits Tech application: They can create large-scale simulations of complex situations such as weather and environmental interactions. They can enable us to understand the long-term consequences of complex events and to improve the design of products or processes from cradle to grave before actually building them.

Local Tech application: They can constantly monitor and manage small-scale situations such as household greenhouse gardens where infections can destroy a small system if not spotted early. Also, they can create local libraries of properly scaled solutions to local problems.

Nature Tech application: They can analyze and catalog genomes of plants and animals.

Human Tech application: They can measure and monitor complex physiological systems (such as hormonal systems) and improve human interactions through analysis of language patterns.

4. Chronobiology: This is the scientific study of time-based patterns in the human body.

Super Tech application: It can be used to optimize human performance during outer-space trips.

Limits Tech application: It can be used to minimize waste of human efforts by putting teams together when they are most efficient at specific kinds of tasks.

Local Tech application: It can be used to match local day and night patterns to human function. There can be very different expectations of human activity in high latitudes where winter days and summer days are of dramatically different lengths compared to tropical locations where lengths of days and nights change very little over a year.

Nature Tech application: Applied to plants and animals, as well as humans, it can help us to find out specific patterns of time-based performance.

Human Tech application: It can help individuals understand and achieve their own optimum performance patterns and to put together teams with the best chronobiological matches.

5. Stereolithographic (3D) Manufacturing: This is the process of using a derivative of ink jet technology to "print" three-dimensional objects by building up very thin layers of material. Instead of ink, the jets squirt out tiny droplets of plastic, metal, or other materials in minute layers, building up complete objects one layer at a time. Research has already demonstrated that a mix of plastic, metal, and ceramic can be incorporated into the same object with this process.

Super Tech application: It can produce a huge range of products on demand in many locations so that everyone has access to almost anything almost instantly. In a very real sense, nothing would ever become obsolete if you had the manufacturing instruction set.

Limits Tech application: It can produce objects with a minimum of waste and with absolutely no overrun, by producing only objects that have been ordered.

Local Tech application: It can manufacture complex objects within the community without having to have expensive manufacturing facilities or importing the object.

Nature Tech application: It can create artificial objects that could not be generated by biological processes.

Human Tech application: It can render into physical form artistic ideas.

6. Hydrogen Fuel: It burns with oxygen to produce mostly H_2O as the primary waste product with small amounts of hydrogen peroxide. It is the most environmentally friendly of burnable fuels as long as it doesn't leak into the atmosphere unburned.

Super Tech application: It's used for all transportation devices because of lightness of fuel, ease of production, and long-term abundance.

Limits Tech application: It's used because its source is not a diminishing resource. It creates the least pollution and does not contribute to CO_2 in the atmosphere.

Local Tech application: It's easily produced in small scale and stored close to the community to be used for local energy and transportation needs.

Nature Tech application: It can be generated with biological processes and used by the community for transportation, heating, and electricity generation.

Human Tech application: No direct use.

7. Holography: This visual display process generates 3D images. It can be done with still images or moving images.

Super Tech application: It's used as the ultimate telephone. The interactions would be between two or more holographic images. It's also great for entertainment and gaming.

Limits Tech application: It's used as part of the simulation process. It can create 3D weather patterns as well as 3D product designs that could be examined for both their esthetics and their functionality before actual production.

Local Tech application: It's used for cataloging local resources and for "illustrating" ideas in education and the arts.

Nature Tech application: It's used to generate 3D genome patterns, protein folding patterns, and cellular structures for research and education.

Human Tech application: It's a theoretical basis for understanding how the human brain stores and accesses memories. It's also a new artistic medium.

8. Lab on a Chip: This technology is made up of a miniature matrix of thousands of individual testing chambers carved onto a clear silicon rubber square of one inch by one inch. It can test up to ten thousand different elements at once or one element ten thousand different ways, thus dramatically speeding up experimentation.

> *Super Tech application:* It can generate a superabundant amount of information on a new material.
> *Limits Tech application:* It can test new chemical formulas on environmentally sensitive agents to make sure that the new chemical passes all tests before being used in the real world.
> *Local Tech application:* It can run very sophisticated pollution tests in local situations at very low cost and with all the analysis done on the spot.
> *Nature Tech application:* It can test the effects of new combinations of naturally occurring enzymes on materials for new uses.
> *Human Tech application:* It can be used to create chips based on an individual's DNA so that tests can be run to indicate the individual's sensitivity to allergens, drugs, foods, and so on.

9. Nanotubes: These carbon-based fibers have extraordinary strength and lightness. Extremely small and capable of conducting electricity with high efficiency, nanotubes can be woven into fabric or assembled into structures of any size.

> *Super Tech application:* They can be used to create aircraft, bridges, and buildings of immense size and strength.
> *Limits Tech application:* They can be used to construct vehicles that would be long lasting (since carbon fibers don't rust), strong (to protect the passengers in an accident), and lightweight (which would reduce the fuel needed to move the vehicle). Another use is as a storage medium for hydrogen gas at room temperature.

Local Tech application: They can be used to construct green-house structures, bicycle frames, and dirigible frames.

Nature Tech application: They can be used for reinforcement of natural structures and for fabrication of elements that cannot be generated through natural processes.

Human Tech application: They can be used as a growth medium and for scaffolding for support after an injury to bone or cartilage.

10. Space Satellites: These orbiting devices can be used for communication, monitoring the planet or outer space, energy generation, and habitation.

Super Tech application: They can be used for bringing communications, both voice and video, to everyone on the planet. In near space they can provide added room for a growing population.

Limits Tech application: They can be used to monitor the health of the planetary ecosystems and development of pollution problems, and watch for planet-killing asteroids.

Local Tech application: They can be used to watch weather patterns and alert local communities to potential threats.

Nature Tech application: They can be used to watch for emerging natural epidemics that could cause damage to ecosystems.

Human Tech application: They can be used to support high levels of communications between various human groups across the planet.

11. Computer Simulations: The ability to generate via computer a representation of some aspect of the real world or an unreal world. We offer only one of hundreds of uses in each region.

Super Tech application: They can be used to create games and activities for entertainment (e.g., *Star Trek's* holodeck).

Limits Tech application: They can be used to teach the concept of feedback loops and systems dynamics to students.

Local Tech application: They can be used to create a model of the local economic system to help determine best policies for sustainable development.

Nature Tech application: They can be used to generate models of cellular functions.

Human Tech application: They can be used to create "artificial" societies within a computer to better understand human behavior.

12. E-book: This electronic device displays text and pictures on a lightweight illuminated thin screen. The number of books it can store is limited only by its memory.

Super Tech application: It can be used to carry and access enormous amounts of information easily.

Limits Tech application: It can be used to replace paper books and thus save precious resources while still providing access to information and entertainment.

Local Tech application: It can be used for paper replacement but also to make it easier for local authors to be "published."

Nature Tech application: The uses are the same as for Limits Tech.

Human Tech application: It allows inexpensive and easy access to the wisdom of the species.

In addition to the many specific examples we have found, there are two enormous categories that are part of the Universal Technologies: biotechnology and nanotechnology. Both of these technologies are very large in their scope and broad in their applications. Neither, by themselves, can create a TechnEcology. But biotechnological and nanotechnological innovations will be very influential in all five regions.

602 ▬

The Best Answer—A Hybrid TechnEcology?

A camel is a horse
designed by a committee.

OLD SAYING

A question we receive almost every time we lecture on the five regions is "Why can't you just take the best technology from each region and make a hybrid?"

We've pondered that question for fifteen years, because it seems like it makes perfect sense. Yet, when you look at the natural world, you don't see hybrid ecosystems. In fact, what you see again and again are clear boundaries between ecosystems. So, a quick answer to the question is that since Mother Nature doesn't do hybrids, and she has been experimenting for several billion years, it probably isn't a good idea.

But there are more satisfying answers than that. One of the issues of any hybrid is the difficulty of passing on the benefits of the hybrid. For instance, to maintain a hybrid plant, you must actively create the hybrid each generation. If you plant the seeds the hybrid produces, the next generation will default back to prehybrid characteristics.

So, there is an ongoing and costly effort that must be maintained in order to continue hybrid benefits.

In human terms, it means that we would have to make an enormous number of decisions about how we want our future to unfold. Each time we wanted to bring in a new technology, we would have to examine how it would interact with all the other technologies we were using, just like adding a new genetic trait has to be examined for how it alters already acceptable traits. This constant analysis of interactions would be required in a hybrid technology world and would consume much time and energy.

The benefit of having a clearly defined TechnEcology is the ease of adding new elements that match the region's values and characteristics. This allows the members of the TechnEcology to judge, with little effort, the fitness of a new technology for their region.

Let us take a look at a project in Japan to illustrate the difficulties of hybridization. Since the late 1990s, Japan has been designing a huge structure called Sky City. It is a single building two thirds of a mile tall that is intended to house one hundred thousand people with the infrastructure necessary for them to live and work within that structure.

We can see elements from all five regions manifested in Sky City.

Super Tech. Consider the sheer size of the undertaking. To build one structure that will house one hundred thousand people, provide them with light, heat, cooling, recreational facilities, and transportation, and be immune to the forces of nature such as earthquakes and typhoons clearly falls within the Super Tech region.

Limits Tech. But, because all the homes share side walls and because the floor for one unit is the ceiling for the one below, building materials are used more efficiently. Because there is much less surface area exposed per dwelling, it will take much less energy to heat and cool the living units. Because everyone will live and work in Sky City, the amount of transportation, public or private, is dramatically reduced, thus saving capital, manufacturing time, materials, fuel, and, of course, travel time. That's the Limits Tech way.

Local Tech. The building is arranged in vertical villages, each of which holds ten thousand people. This matches a Local Tech value and creates conditions for local cooperation that are rare in standard Super Tech cities.

Nature Tech. Sky City has a series of open parks at different levels. This creates an opportunity to have living things, both plants and animals, become an integral part of the structure. And the immense amount of windows invites growing things inside the dwelling and perhaps even on the outside.

Human Tech. How do you create architecture that fits one hundred thousand people into such a small space and still have a civilized society? That would be one of the key focal points for the Human Techers in Sky City: to design new human interaction networks where vertical travel is more important than horizontal travel.

So, important facets of all five regions can be found in Sky City. But, let's apply a simple test: Who do you believe would be the most comfortable in Sky City? We think only Super Techers would find it well suited to their values.

This is the dilemma with hybrids. By taking one item from column A and two items from column B, and so on, the designers construct a technological system that works in the ideal, but fails when human values are applied.

On a weekly basis, we see Super Tech organizations trying to solve a problem they have created by applying Limits Tech solutions. One example from Ford Motor Company will illustrate this.

In 2002 Ford announced that they were going to build an even bigger F150 Ford Truck. Bigger equals Super Tech. But, since they were being criticized for the poor mileage their trucks got, they announced several months later that they were going to build a hybrid system to increase the fuel mileage of their new truck by 30 percent. That was a Limits Tech solution being applied to Super Tech to fix the problem the bigger truck was creating. Ford is trying to deal with the value differences between the two regions. And it is very challenging.

In many ways, the United States has been trying to hybridize. We believe it is one of the main reasons that the nation is struggling to find its vision of the future. And that struggle will continue until we decide on the dominant values we want to use to inform our technological decisions. One more example: In response to the criticism starting in the 1960s, Super Techers said they would eliminate pollution, save resources, increase car mileage, conserve energy, and protect the environment. Yet, instead of continually improving in those areas, every time we came to a place where we could choose between "bigger is better" versus "smaller is sustainable," we chose bigger. Look at our SUVs. In less than a decade, we, as a nation, killed all the mileage gains we had worked on for twenty years. Look at our McMansions. Even empty nesters have felt the need for five-thousand–square-foot homes. Look at our two-home families and our four-car families. Most of those Americans who had enough money to make a choice chose Super Tech.

Our Super TechnEcology has tried alternatives, but the basic values remain and therefore we keep returning to the norm. And those values put Super Tech at odds with the Limits, Local, and Nature TechnEcologies.

Those three are structured to sustain a kind of civilization where it is easy to co-exist, long term, with the natural world. Super Tech just doesn't have ways of doing that. So, the Super TechnEcology must find a place where its values do not threaten the planet. And we believe there is such a place waiting.

Here is how we think the combination of sustainability and Super Tech could work.

THE EARTH-SUSTAINING REGIONS: LIMITS, LOCAL, NATURE AND HUMAN TECH

We believe that Limits, Local, Nature, and Human Tech can produce sustainable futures without damaging the natural ecosystems of Earth. Limits Tech is the quintessential version of sustainability with its willingness to shrink human population and slow down change to reinforce the concept.

Local Tech achieves sustainability through molding itself to local conditions and staying small enough so that it never pushes at the natural boundaries too hard. Its emphasis on being a shepherd of Nature, protecting her while harvesting what we need, translates into a desire to preserve and maintain, which it can do in any set of local conditions.

And, obviously, Nature Tech marries us to the natural life processes to take advantage of the lessons of evolution we have learned over the millennia.

Human Tech starts from inside us and works out. Its focus can sustain us anywhere because the technology most important to us has been inside us all the time.

But none of these regions has the unique capacity to protect the entire planet from threats beyond. Only Super Tech has the values that allow it to succeed in such a role.

SUPER TECH IN OUTER SPACE

We think that the Super TechnEcology has to move off-planet. It can have several roles there. First and foremost, Super Techers will staff the watchtowers—they will be sentinels looking for danger, whether this danger is a stray comet or a planet-killing asteroid. We now know that there have been several mass extinctions on Earth caused by asteroids or comets. It has happened every sixty million years or so with the last one about sixty-five million years ago.

Being the sentinel and protector is an essential role for any species that plans to be around a long time. Some have suggested that humankind's quest for flight since time immemorial is a primal response to this cosmic danger. To do this well would also require cleaning up the Solar System. By removing solar debris such as Earth-orbit–intersecting asteroids, Super Techers eliminate threats as well as provide additional resources for their expansion into the Solar System.

Super Tech's outward focus keeps our species from being too introverted and Earth-centered. We have expanded across the planet, starting from one small place. Had our ancestors decided to stay there and never leave, where would we be now?

Our species has been on a 150,000-year odyssey. Unfortunately, our success has left us full of hubris, thinking that upward and onward, bigger and better can solve all problems. But when we leave the planet, we are immediately humbled by seeing the small place we have in the universe. Humbled but striving is the attitude of Super Techers in outer space.

Outer space is the "right" environment for the Super Techn-Ecology. In that domain, it can support the planet and the other regions, and it can generate new options for colonization rather than destroying Mother Earth. This is the special role of the Super TechnEcology.

603
"Tyger, Tyger, burning bright . . ."

We are living in a time where the technologies that underpin the design and structure of our societies around the world are in tremendous ferment. A key indicator of this turbulence, as we have tried to illustrate throughout this book, is the number of solutions that are available to solve the major problems of the world. We hope we have proved the claim we made in the preface: *We are at one of those rare times in history when we have more solutions than we have problems.*

Not only do we have a surplus of solutions, but those solutions are clustering into a new kind of structure—TechnEcologies. And, like their biological cousins, these systems are developing the capacity to adapt, evolve, and expand.

In addition, since these technological ecosystems are only beginning to emerge, humanity has time to examine and understand how to deal with these new systems. We have failed to do that in the past, and the results, repeatedly, have been surprises, unintended consequences, and worldwide change, some of which has been good and some of which has been very bad.

It is worth looking at the emergence of the automobile again and adding some additional details to our discussion from the opening chapter, to illustrate what we mean.

A CAUTIONARY TALE OF UNINTENDED CONSEQUENCES

At the turn of the twentieth century the automobile was introduced as a new technology. It quickly came to dominate the era with profound consequences for society.

The automobile needed reliable power and was soon allied with gasoline as its major fuel. This spawned the growth of two large industries: autos and oil.

There were massive consequences of this victory of automobiles and oil over both the horse-and-buggy technology and alternative technologies that were competitors at the turn of the twentieth century. Few of these consequences were anticipated or discussed at the time. But they were far reaching. Here are a few of them.

With the growth in demand for oil came a global search for additional supplies. This led to national relationships based on oil exploration and exploitation. Because these sources were on other continents, it led to the development of enormous oil tankers to bring oil to the major markets. The needs of this global supply chain influenced geopolitics, the creation of alliances, and the development of armies and navies to protect them.

Setting up local distribution networks created another set of social and economic changes. Building refineries, pipelines, storage tanks, truck tankers, and gas stations changed the landscape (and smell) of cities, towns, and even remote rural areas around the world.

The autos-and-oil lifestyle that developed has had long-term effects on foreign policy and world events. Areas of the world that are rich in oil, from the Middle East to Alaska, have become powerful. Foreign alliances and wars have centered on the need to provide and protect the oil supply—largely to continue to supply the automobile's need for gasoline.

An ensemble of technologies grew around the car, with the steel and rubber industries also developing to meet the demand for more automobiles. Even though the automobile was not a whole Techn-Ecology by itself, several products and even industries like oil, steel, and rubber acquired a common interest in the ongoing success of the auto.

Once large enough, the combination of autos and oil became a predator in the larger technological ecosystem. Potential competitors or alternatives to autos and oil were eliminated. The auto companies themselves bought successful urban electric trolley systems and replaced them with gasoline buses. Touring cars replaced once-common passenger trains. Even a good portion of the rail freight traffic has been taken over by semitrailer trucks.

Cars also needed a relatively smooth surface to drive on so streets and highways were built. Governments used its power to acquire land, even over the objection of private landowners, for this ever-expanding network of roads.

The subsequent improvement in roads and highways and the increase in the speed of cars allowed people to live farther from work, which led to urban sprawl and the birth of suburbs.

Once people could quickly and easily drive long distances, services and the distribution of goods could be concentrated in enormous supermarkets and malls. One consequence of this development was the "malling" of America and the disappearance of the corner store from most cities and towns.

Other social changes occurred as well. Courting habits and entertainment options changed. Young men did not call on young women at their home to go for a walk anymore. They drove off in the car, and every little town had a road that was known officially or unofficially as "lovers' lane."

People could travel long distances for all forms of entertainment. Fans from a large geographic area could support professional sports teams, theater, and concerts. Family travel via automobile across the country became the major mode of vacations.

As the auto industry concentrated in Detroit, cars came to define the personality of the city and the region. Cars began to dramatically shape our modern civilization. They were the get-to-work vehicle and also the leisure vehicle leading to drive-in theaters, root beer stands, and fast-food chains where you don't even have to leave the car. Automobiles have become personal space with an expectation that we have personal choice with options of individual travel, even at the expense of one of the greatest problems of the age—pollution.

At the beginning of the Auto Age we didn't even have a concept for pollution. It came as a long-term, unintended consequence. But it's now a major shaper of events relating to our world's environment. And the single greatest polluting agent on the planet is the gasoline-powered automobile.

Obviously, we could continue the litany of consequences triggered by the gasoline-powered automobile. The point is simple: When the car was first introduced into our society, we did not explore the possible implications. Instead we lived them out and ended up with problems that could have been avoided, had we taken the time at the beginning to find out what we were getting into.

The Exclusionary Principle

There is one more crucial lesson to be taken from the automobile saga. Many have forgotten that the gasoline-powered engine was only one of four kinds of power sources competing for this new market. There was also the Stanley Steamer, which burned its fuel externally to heat up the boiler; the first electric car, which used batteries and could be recharged at home; and Daimler's diesel engine, which used a low grade of fuel and ignited that fuel with high-compression forces instead of a spark plug.

Why don't we have equal amounts of these technologies today? Why did the gasoline engine become almost the sole solution for automobiles?

The pattern is simple and can be seen again and again in the adoption of new technologies. Initially all four systems had fairly equal footing. Some technology historians argue that the Stanley Steamer was actually the best technology at that time. But little things can cause enormous swings. As cars became more popular, horse transportation became less so. Since there were fewer horses being used for transportation, watering troughs for the horses began to be removed. Guess where the Stanley Steamer got its water.

At the same time, gasoline-powered cars got a huge lift because someone named Henry Ford decided to build a very inexpensive car for the masses (even though the automobile was considered a toy of the rich). He picked the gasoline engine for his Model A.

Once the momentum began to build for gas engines, entrepreneurs began to figure out ways to get on that bandwagon: better ways to make gasoline from the oil companies; better ways to pump gasoline for the fuel stations; better spark plugs to fire the ignition; better electric coils to make stronger sparks. Notice that there is only one thing on this list that would have benefited the other three competing technologies.

In the larger scheme, investment money began flowing to try to take advantage of the success of the gas-powered car. And investment in the other technologies began to slow down.

Within a decade, the gasoline engine was the technology to use for any automobile. The rest is history.

So, success draws investment, which funds more success. This is a "virtuous" spiral for the winning technology and a "death" spiral for the competing technologies.

We could have just as easily illustrated the exclusionary principle with Microsoft or Intel or Starbucks.

This same pattern may very well affect the five regions of the future. For instance, if scientists are able to create a fusion reactor to generate electricity, what happens to the interest in developing aerogel insulation for homes? Why insulate when you have all the cheap energy you need to heat and cool your home? That's how the exclusionary principle works.

Why develop windmill technology or wave power or hydrogen-gas–producing bacteria if you already have fusion power? Who would invest their money in a project that has no significant competitive advantage?

And, if we don't openly discuss and explore our options for the next hundred years, then without our control, without our influence, the exclusionary principle will wield its power to shape our future.

Could the exclusionary principle lead to fewer regions or the dominance of one? We think it could happen but that it would be a great loss.

How Many Regions Do We Need?

So, does the exclusionary principle imply that we will have only one of the five regions by 2100?

No. But it suggests that it is possible if we do not actively participate in preventing it. What's wrong with only one? Having multiple TechnEcologies on Planet Earth is a way to guarantee that we have options.

In fact, if technological ecosystems follow the pattern of biological ecosystems, there will be many more than five by the end of the twenty-first century. Mother Nature has more than one hundred ecosystems on the planet operating right now. Thousands probably have come and gone during the history of the Earth.

There is another extremely important reason not to end up with just one TechnEcology. Research on innovation in living systems has discovered that most of the radical innovations in Nature occur at a verge.

THE VERGE

A "verge" is where two different things meet, such as the edge where two or more ecosystems come together. The reason that this area is rich in innovation is simple: Only at the verge can you encounter something that is totally outside of your experience. This novel experience triggers new responses from old elements, causing them to mutate to take advantage of or respond to the new situation. It also invites the combining of elements from different systems to create new partnerships for dealing with the world. Close proximity also makes it much easier to exchange genetic material, which happens between plants on a regular basis.

Only with multiple TechnEcologies can you have a wealth of verges. And the verges will create ongoing opportunities for innovation. This system has worked for Mother Nature for more than three billion years. It would be foolish of us not to learn from her.

USING THE FIVE-REGIONS IDEA

These five categories give us a more precise way of looking at and defining our new technologies. For both public and private discussions, these categories can help frame the question: On which region or regions should we focus our energy and resources?

For public policy, these categories help us measure what we are regulating and funding. Right now, governments fund "high technology." That's not discriminating enough any more.

For various parts of the world, the five-regions model becomes a rallying point for designing the future. Each region is the seed for a vision of the future.

The real challenge is how to create an ongoing public dialog across the planet to address the issues that are generated by these extraordinary times.

FIVE REGIONS DEVELOPMENT

Is there a place where you can go to find each of the five regions of the future? Not yet. We think that there are groups at the crossroads between TechnEcologies. No nation has totally committed to any one direction. Even so, it looks like the USA and Japan are leading the way to the Super Tech region and Denmark seems to be moving toward Local Tech. There have been reports that the newly emerging Afghanistan is interested in the village life of Local Tech as well. Most other countries, once they examine the characteristics of each of the five regions, will probably find one that best matches their own values.

How do we think the situation will unfold? Nations and even regions (groups of small nations or parts of very large nations) will embrace specific TechnEcologies over the next fifteen to twenty years.

In the twenty-first century, we need a more sophisticated way to catalog and describe our technology. We think the five regions offer that. As citizens of this new world, we all need to begin to think more systemically. The five-regions methodology invites that. Our technologies are bigger than our nations. We need to understand the consequences of that.

In every age, technology has been a major shaping force. So it will be in the twenty-first century. It is time to get ready for that.

Two hundred years ago, the poet William Blake wrote a poem about technology. His metaphor for technology was the tiger.

> *Tyger, tyger, burning bright*
> *In the forests of the night,*
> *What immortal hand or eye*
> *Could frame thy fearful symmetry?*

Blake understood two hundred years ago the power and challenge of technology. Now, we must try to answer his questions in our own era. The level of complexity is vastly greater now than in Blake's time. And it is vastly more powerful. Even as his generation grappled with the tiger, trying to make sense of it, trying to cage it so that it did not destroy humanity, so must our generation. We have been creating an inventory of tools that has the potential to make a world better than we have ever dreamed. But those same tools can also be used for vast harm.

So the dialog must begin—not just between elites and technocrats, but with everyone. It must be a wide-open, public discussion. We must illuminate the forests of the night. We must explore the long-term implications of our emerging TechnEcologies.

Now, it is our turn to frame the fearful symmetry.

Bibliography

Introduction

001 A New Look at Technology

Barker, Joel A. *Future Edge: Discovering the New Paradigms of Success.* New York: William Morrow, 1992.

Barker, Joel A. *Paradigms: The Business of Discovering the Future.* New York: HarperBusiness, 1993.

Christensen, Clayton M., Scott D. Anthony, and Erik A. Roth. *Seeing What's Next.* Boston: Harvard Business School Press, 2004.

Christensen, Clayton M., and Michael E. Raynor. *The Innovator's Dilemma.* Boston: Harvard Business School Press, 1997.

Christensen, Clayton M., and Michael E. Raynor. *The Innovator's Solution.* Boston: Harvard Business School Press, 2003.

Naisbitt, John. *Megatrends.* New York: Warner Books, 1982.

004 Your Awareness of TechnEcologies

Schwartz, Peter. *The Art of the Long View.* New York: Currency Doubleday, 1996.

Chapter 1 Super Tech

101 Overview and Guidelines

Hoffman, Carl. "Conspicuous Construction." *Popular Science,* April 2004, pp. 50–58.

102 Advocates and Examples

Abelson, Philip H. "Avoiding an Oil Crunch." *Science,* October 1, 1999, p. 47.

Artificial heart,
http://www.heartpioneers.com/

Artificial organs,
http://www.bae.ncsu.edu/research/blanchard/www/465/textbook/otherprojects/2002/group_A/menu.html
http://www.blackwellpublishing.com/journal.asp?ref=0160-564X

"Artificial Retina." *Technology Review,* September 2004, pp. 82–86.

Asimov, Isaac. *I, Robot.* New York: Bantam Spectra Books, 1950.

"Asteroid Threat Is Greater than Ever." *New Scientist,* September 14, 2002, p. 6.

"The Best Goes On." *Discover,* January 2000, p. 33.

"Bionic Connection." *Discover,* November 2002, p. 49.

Boyd, Philip. "Ironing Out Algal Issues in the Southern Ocean." *Science,* April 16, 2004, p. 396.

Clarke, Arthur C. *2001: A Space Odyssey.* New York: Roc, 1968.

Doctorow, Cory. "Rise of the Machines." *Wired,* July 2004, pp. 115–119.

Edwards, Cliff. "Ready to Buy a Home Robot?" *Business Week,* July 19, 2004, pp. 84–90.

"Extra Arms for Heart Surgeons." *Business Week,* November 22, 1999, p. 67.

Fairley, Peter. "Saving Lives with Living Machines." *Technology Review,* July/August 2003, pp. 59–63.

Fallows, James. *Free Flight.* New York: Public Affairs, 2001.

Fiber optics,
http://electronics.howstuffworks.com/fiberoptic.htm

Fuller, R. Buckminster. *Operating Manual for Spaceship Earth.* Mattituck, NY: Aeonian Press, 1976.

Fusion Energy Education Site,
http://fusedweb.pppl.gov/

"Gentlemen, Start Your Robots." *Technology Review,* December 2002, p. 31.

Goldman, Lea. "Machine Dreams." *Forbes,* May 27, 2002, pp. 149–150.

Graham-Rowe, Duncan. "We Can Rebuild Them . . ." *New Scientist,* February 28, 2004, pp. 26–29.

Hauge, Frederic, and Marius Holm. "Give Carbon a Decent Burial." *New Scientist,* July 17, 2004, p. 16.

Heinlein, Robert. *The Past through Tomorrow.* New York: Putnam, 1967.

Horvath, Joan C. "Blastoffs on a Budget." *Scientific American,* April 2004, pp. 92–97.

Hudson, Gary C. "The ROTON Concept and Its Unique Operations." http://www.spacefuture.com/archive/the_roton_concept_and_its_unique_operations.shtml

Kahn, Herman. *The Coming Boom*. New York: Simon & Schuster, 1982.

Kahn, Herman, Leon Martel, and William Brown. *The Next 200 Years*. New York: Harper, 1976.

Kahn, Herman, and Anthony J. Wiener. *The Year 2000: A Framework for Speculation on the Next 33 Years*. New York: Macmillan, 1967.

Koerner, Brenden I. "The Trillion-Barrel Tar Pit." *Wired*, July 2004, pp. 102–104.

"Laser-Guided." *The Economist*, March 25, 2000, p. 66.

Lomborg, Bjørn. *The Skeptical Environmentalist*. Cambridge, UK: Cambridge University Press, 2001.

Mann, Charles C. "The Bluewater Revolution." *Wired*, May 2004, pp. 183–187.

May, Mike. "When Jeep Meets Jump-Jet." *New Scientist*, June 14, 2003, pp. 40–43.

Megacities, http://mc2000.arch.hku.hk/megacities_now.pdf

Molitor, Graham T.T. "The Next 1,000 Years: The 'Big Five' Engines of Economic Growth." *Futurist*, December 1999, pp. 13–19.

Niven, Larry. *Ringworld*. New York: Ballantine, 1970.

O'Leary, Brian. *The Fertile Stars*. New York: Everest House, 1981.

O'Neill, Gerard K. *The High Frontier*. New York: William Morrow, 1977.

Port, Otis. "More Bang from the Bubble?" *Business Week*, March 29, 2004, pp. 84–85.

"Priming the Pump." *Red Herring*, October 2002, p. 30.

Robots, http://www.aaai.org/AITopics/html/robots.html

"Robots Dig In to Defend Earth." *Popular Mechanics*, August 2004, p. 19.

Schmitt, Harrison H. "Mining the Moon." *Popular Mechanics*, October 2004, pp. 56–63.

"Seeing-Eye Chip." *Forbes*, October 14, 2002, p. 160.

Simon, Julian. *The Ultimate Resource 2*. Princeton, NJ: Princeton University Press, 1996.

Space 2100: To Mars and Beyond in the Century to Come. New York: Time Inc. Home Entertainment, 2003.

"Speak, Aibo, Speak!" *Popular Science*, August 2004, p. 80.

Spencer, John. *Space Tourism*. New York: Apogee Books, 2004.

Stapledon, William Olaf. *Last and First Men and Star Maker*. New York: Dover, 1968.

Stover, Dawn. "Turn the ISS into a $2 Million-a-Night Space Hotel." *Popular Science*, May 2004, pp. 81–90.

Sweetman, Bill. "Whoosh." *Popular Science*, July 2004, pp. 56–62.

"Targeted Therapies." *Science News*, September 14, 2002, p. 171.

"A Vision of the Future." *Scientific American*, October 2002, p. 51.

"Wearable Machines." *Popular Science,* July 2004, p. 40.

"Wearable Robots." *Technology Review,* July/August 2004, pp. 70–73.

Weiss, Rick. "Mind Over Matter: Brain Waves Guide a Cursor's Path." *washingtonpost.com,* http://www.washingtonpost.com/wp-dyn/articles/A59791-2004Dec12.html

Wilson, Jim. "Dangerous Science." *Popular Mechanics,* August 2004, pp. 75–79.

Wolman, David. "Hydrates, Hydrates Everywhere." *Discover,* October 2004, pp. 62–67.

Chapter 2 Limits Tech

201 Overview and Guidelines

Carson, Rachel. *Silent Spring.* Boston: Houghton Mifflin, 1962.

Meadows, Donella H., et al. *The Limits to Growth.* New York: Universe Books, 1972.

202 Advocates and Examples

Anti-antibiotics, http://archives.cnn.com/2000/HEALTH/06/01antibiotic.overuse

Birth control, www.populationconnection.org

Books, www.ebooksys.com

"The Bright Idea That Could Replace Your Flickering Fluorescents with a Cool Glow." *New Scientist,* August 3, 2002, p. 16.

Buechner, Maryanne Murray, Lev Grossman, and Anita Hamilton. "Coolest Inventions 2002." *Time,* November 18, 2002.

Callenbach, Ernest. *Ecotopia.* New York: Bantam, 1990.

Carbon fiber, www.psrc.usm.edu/macrog/carfib.htm

Carson, Rachel. *Silent Spring.* Boston: Houghton Mifflin, 1962.

Computer Simulations, www.fhwa.dot.gov/reports/pittd/computer.htm www. newsmedical.net/?id=4132

Co-products, www.acaa-usa.org

Daly, Herman E. *Beyond Growth.* Boston: Beacon, 1997.

Diamond, Jared. *Collapse: How Societies Choose to Fail or Succeed.* New York: Viking, 2004.

Diamond, Jared. *Guns, Germs and Steel.* New York: W.W. Norton, 1999.

E-Books, www.booksys.com

Erhlich, Paul. *The Machinery of Nature*. New York: Simon & Schuster, 1986.

Erhlich, Paul. *The Population Bomb*. New York: Ballantine, 1968.

Fishing, http://archive.greenpeace.org/oceans/globaloverfishing/

Food growing, www.precisionfarming.org

"Goodbye, Mister Edison." *Discover*, November 2002, p. 18.

Hardin, Garrett. "Tragedy of the Commons." *Science*, 162 (1968), pp. 1243–1248.

Hawken, Paul. *The Ecology of Commerce*. New York: HarperBusiness, 1993.

Hawken, Paul, Amory Lovins, and L. Hunter Lovins. *Natural Capitalism: Creating the Next Industrial Revolution*. Boston: Little, Brown, 1999.

"High Speed Copper Wire Communications," *IEEE Journal*, December 1995, pp. 1571–1585, www.electronicstalk.com/news/not/not103.html

"LEDS Light the Future." *Technology Review*, September/October 2000, p. 39.

Levin, Simon. *Fragile Dominion*. New York: Perseus, 2000.

Lovins, Amory B. "Energy Strategy: The Road Not Taken?" *Foreign Affairs*, October 1976, pp. 65–96.

McKibben, Bill. *Hope, Human & Wild*. Boston: Little, Brown, 1995.

Meadows, Donella H., et al. *Beyond the Limits to Growth*. White River Jct., VT: Chelsea Green, 1993.

Meadows, Donella H., et al. *The Limits to Growth*. New York: Universe Books, 1972.

Meadows, Donella H., et al. *Limits to Growth: The Thirty Year Update*. White River Jct., VT: Chelsea Green, 2004.

Nattrass, Brian, and Mary Altomare. *The Natural Step for Business*. New York: New Society, 1998.

Pimm, Stuart L. *The World according to Pimm*. Boston: McGraw-Hill, 2001.

Quinn, Daniel. *Beyond Civilization*. New York: Three Rivers, 1999.

Quinn, Daniel. *Ishmael*. New York: Bantam, 1992.

Quinn, Daniel. *The Story of B*. New York: Bantam, 1996.

"Radiation for Food Preservation." *Frontline*, August 16–29, 2003, www.frontlineonnet.com/f12017/stories

Rees, Martin. *Our Final Hour*. New York: Basic Books, 2003.

Riley, Robert Q. *Alternative Cars in the 21st Century*, 2nd ed. Society of Automotive Engineers, 2003, www.rqriley.com/alt=car2.htm

Satellites, www.satnews.com/microsatellite.html

"Super Conducting Cable to Increase US Grid Density." *Nexans Company News*, April 30, 2003.

Superinsulation of homes, Montana Super Insulation Project, www.nahn.com/mtsiproj.htm

Thermal conversion process, www.changingworldtech.com
Trains, www.transportation.anl.gov/research/materials/laser_glazing.html
   www.sandia.gov/media/newsRel/nr2000/seraph.htm

Chapter 3 Local Tech

301 Overview and Guidelines

Schumacher, E.F. *Small Is Beautiful*. New York: Harper & Row, 1973.

302 Advocates and Examples

Airships, www.aboutairships.com
Anderson, Claire. "Southern Comfort in a Straw Bale Home." *Mother Earth News,* June/July 2004, pp. 50–57.
Ashley, Steven. "Fuel Cell Phones." *Scientific American,* July 2001, p. 25.
Brady, Diane. "Reaping the Wind." *Business Week,* October 11, 2004, pp. 201–202.
Brower, Kenneth. *The Starship and the Canoe*. New York: Bantam, 1979.
Brown, Lester R. *Building a Sustainable Society*. New York: W.W. Norton, 1982.
Brown, Lester R. *Eco-Economy: Building an Economy for the Earth*. New York: W.W. Norton, 2001.
Brown, Lester R. *Plan B: Rescuing a Planet under Stress and a Civilization in Trouble*. New York: W.W. Norton, 2003.
Brown, Lester R. "Turning on Renewable Energy." *Mother Earth News,* April/May 2004, pp. 100–107.
"Catalyst Boosts Hopes for Hydrogen Bonanza." *Science,* September 27, 2002, p. 2189.
Chiras, Dan. "Building with Earth." *Mother Earth News,* April/May 2002, pp. 26–35.
Creedon, Jeremiah. "Increase Your Energy IQ." *Utne Reader,* July/August 2004, pp. 86–87.
Davis, Katharine. "Enough Wind to Power the World." *New Scientist,* September 25, 2004, p. 12.
Elgin, Duane. *Voluntary Simplicity*. New York: William Morrow, 1993.
Expo 2005 Aichi, Japan, www-1.expo2005.or.jp/en/
Fairley, Peter. "Solar-Cell Rollout." *Technology Review,* July/August 2004, pp. 35–40.
Foertsch, Catie. "Electrifying Flight." *Air & Space,* October/November 2004, p. 13.

Food, www.localharvest.org, www.greenhouse.com

Fritsche, Olaf. "Dirigibles to Grace Skies over Germany Once Again." *Science,* February 9, 2001, pp. 973–974.

Goho, A. "Special Treatment: Fuel Cell Draws Energy from Waste." *Science News,* March 13, 2004, p. 165.

Halweil, Brian. "The Argument for Local Food." *World Watch,* May/June 2003, pp. 20–29.

Henderson, Hazel. *Beyond Globalization.* Bloomfield, CT: Kumarian Press, 1999.

Henderson, Hazel. *Building a Win–Win World.* San Francisco: Berrett-Koehler, 1996.

Henderson, Hazel. *Creating Alternative Futures.* Bloomfield, CT: Kumarian Press, 1996.

Henderson, Hazel. *Paradigms in Progress.* San Francisco: Berrett-Koehler, 1995.

Hogan, Jenny. "Solar Power Set for Take-off." *New Scientist,* June 7, 2003, p. 14.

Holthusen, T. Lance, ed. *The potential of earth-sheltered and underground space: today's resource for tomorrow's space and energy viability: proceedings of the Underground Space Conference and Exposition, Kansas City, Mo., June 8–10, 1981.* New York: Pergamon Press, 1981.

Hürter, Tobias. "GE: Green and European." *Technology Review,* September 2004, pp. 24–25.

"It Takes Tech to Tango." *Popular Science,* May 2003, p. 82.

Light rail, www.lrta.org

Mandelbaum, Robb. "Denmark Is Working Out Just Fine." *Discover,* June 2004, pp. 48–55.

McDaniel, Andi. "Have Hoe, Will Travel." *Utne Reader,* May/June 2004, pp. 85–86.

Millett, Stephen M. "Personalized Energy." *The Futurist,* July/August 2004, pp. 44–48.

Moench, Mel. "Self-Sufficient Homes." *The Futurist,* May/June 2004, pp. 45–50.

Morris, David. See Institute for Local Self-Reliance, http://www.ilsr.org/

"Never Thirsty Again." *Economist,* May 31, 2003, p. 42.

Nowak, Rachel. "Power Tower." *New Scientist,* July 31, 2004, pp. 42–45.

Photovoltaics, www.solarbuzz.com/StatsCosts.htm

Port, Otis. "Another Dawn for Solar Power." *Business Week,* September 6, 2004, pp. 94–95.

"Power from Thin Air." *Popular Mechanics,* September 2004, p. 17.

"Power of the Midday Sun." *New Scientist,* April 10, 2004, p. 26.

Quinn, Linda J. "Why Whole Foods Are Better." *Mother Earth News,* June/July 2004, p. 18.

Register, Richard. *Ecocities.* Berkeley, CA: Berkeley Hills Books, 2001.

Run-of-the-River hydro, www.ineed2know.org/HydroPower.htm

Sailboats, www.solarsailor.com

Schumacher, E.F. *A Guide for the Perplexed*. New York: Harper & Row, 1978.

Schumacher, E.F. *Small Is Beautiful*. New York: Harper & Row, 1973.

Segway, www.segway.com

Sky stations, www.astron.nl/craf/skystat/htm

"Solar on the Cheap." *Technology Review*, January/February 2002, p. 50.

"The Solid Future of Rapid Prototyping." *The Economist Technology Quarterly*, March 24, 2001, pp. 49–51.

Staedter, Tracy. "Wave Power." *Technology Review*, January/February 2002, p. 86.

Stereolithography, www.versalaser.com

"Tapping the Tides." *Technology Review*, May 2003, p. 24.

Tilford, Gregory L. *Edible and Medicinal Plants of the West*. Missoula, MT: Mountain Press, 1997.

TWIKE, http://www.phlogma.com/aporia/twike/twikerp.htm

Varchaver, Nicholas. "How to Kick the Oil Habit." *Fortune*, August 23, 2004, pp. 101–114.

Waterless toilet, www.thenaturalhome.com/clivusmultrum.htm

"Wind Power for Homes." *The Futurist*, July/August 2004, p. 2.

Chapter 4 Nature Tech

401 Overview and Guidelines

Watson, James, and Francis Crick. "A Structure for Deoxyribose Nucleic Acid." *Nature*, April 2, 1953, p. 737.

402 Advocates and Examples

"Algae Hydrogen Harnessed." *The Christian Science Monitor*, February 24, 2000, p. 14.

"Animal Self-Medication." *The Ecologist*, March 22, 2002.

Ashley, Steven. "Bumpy Flying." *Scientific American*, August 2004, pp. 18–20.

Benyus, Janine M. *Biomimicry: Innovation Inspired by Nature*. New York: Quill, 1998.

Brown, Kathryn. "Plants 'Speak' Using Versatile Volatiles." *Science*, June 28, 2002, p. 2329.

Coghlan, Andy. "Bugs Broadcasting Corporation." *New Scientist,* April 13, 2002, pp. 12–13.

"Creepy-Crawly Care." *Science News,* October 23, 2004, pp. 266–269.

Dayton, Leigh. "Philodendrons Like It Hot and Heavy." *Science,* April 13, 2001, p. 186.

"Fly Lends an Ear to Microphone Design." *Science News,* December 8, 2001, p. 359.

Forristal, Linda Joyce. "Michael Zasloff: Antibiotic Sleuth," Article No. 13874, Section: Natural Science, December 1988, www.worldandi college.com

Gartner, John. "Algae: Power Plant of the Future." *Wired News,* August 19, 2002, www.wired.com/news/technology/0,1282,54456,00.html

Goreau, T.J. and W. Hilbertz, Coral Reefs, http://globalcoral.org/reef_restoration_projects.htm

Graham, Sarah. "Bacterial Battery Converts Sugar into Electricity." *Scientific American,* September 8, 2003, www.sciam.com/article.cfm?articleID=00054931-E680-1F58-905980A84189EEDF&sc=I100322

Jacobs, Jane. *The Nature of Economies.* New York: Vintage, 2001.

Lovelock, James E. *The Ages of Gaia.* New York: W.W. Norton, 1988.

Margulis, Lynn, and Dorian Sagan. *Microcosmos: 4 Billion Years of Evolution from Our Microcrobial Ancestors.* Berkeley: University of California Press, 1997.

McDonough, William, and Michael Braungart. *Cradle to Cradle: Remaking the Way We Make Things.* New York: North Point Press, 2002.

"Microlenses." *Nature Science Update,* August 2001, www.nature.com/nsu/010823-11.html

"Molecular Computing." *Technology Review,* May/June 2000, pp. 70–77, 81–84.

Ormerod, Paul. *Butterfly Economics.* New York: Pantheon, 1998.

Panchak, Patricia. "Molecular Electronics." *Industry Week,* December 1, 2002.

"Plant Power." *Technology Review,* September 2004, p. 33.

"Plastic from Plants." *Popular Science,* April 2000, p. 22.

Randerson, James. "Bug Clears Arteries." *New Scientist,* October 5, 2002, p. 19.

"Reef Therapy." *Science,* September 3, 2004, p. 1398.

Rothschild, Michael L. *Bionomics: Economy as Ecosystem.* New York: Henry Holt, 1992.

Service, Robert F. "Mammalian Cells Spin a Spidery New Yarn." *Science,* January 18, 2002, pp. 419–421.

"Soaking Up Rays." *Science News,* August 4, 2001, p. 77.

"This Milk Will Make You Grow." *New Scientist,* January 8, 2005, p. 15.

Watson, James, and Francis Crick. "A Structure for Deoxyribose Nucleic Acid." *Nature,* April 2, 1953, p. 737.

Chapter 5 Human Tech

501 Overview and Guidelines

Aristotle. *Nicomachean Ethics*. New York: Oxford University Press, 1998.

Locke, John. *An Essay Concerning Human Understanding*. New York: Prometheus Books, 1994.

Skinner, B.F. *Walden II*. Upper Saddle River, NJ: Prentice Hall, 1976.

502 Advocates and Examples

Adler, Alfred. *Understanding Human Nature*. Center City, MN: Hazelden, 1998.

Arnst, Catherine. "I Can See Clearly Now." *Business Week,* July 31, 2000, pp. 114–116.

Barnett, Adrian. "Fair Enough." *New Scientist,* October 12, 2002, pp. 34–37.

Blanchard, Kenneth and Spencer Johnson. *The One Minute Manager*. New York: William Morrow, 1982.

Blüher, Matthias, Barbara B. Kahn, and C. Ronald Kahn. "Extended Longevity in Mice Lacking the Insulin Receptor in Adipose Tissue." *Science,* January 24, 2003, pp. 572–574.

Bourchard, Thomas, www.psych.umn.edu/psylabs/mtfs/special/htm

"Breast Milk Component Assails Rotavirus." *Science News,* May 16, 1998, p. 317.

Bridges, William. *Managing Transitions: Making the Most of Change*. New York: Perseus, 2003.

Christensen, Damaris. "Medicinal Mimicry: Sometimes, Placebos Work—but How?" *Science News,* February 3, 2001, p. 74.

Chronobiology, http://www.msi.umn.edu/~halberg/introd/index.html

Covey, Steven. *The 8th Habit*. New York: Free Press, 2004.

Deming, W. Edwards, www.deming.org

"Depression Proves Risky for Ill Hearts." *Science News,* February 13, 1999, p. 102.

Drucker, Peter, www.pfdf.org

Ferguson, Marilyn. *The Aquarian Conspiracy*. New York: Jeremy P. Tarcher, 1987.

Frankl, Victor. *Man's Search for Meaning*. New York: Pocket, 1997.

Freud, Sigmund. *The Ego and the Id.* New York: W.W. Norton, 1960.

Frey, William, and Muriel Langseth. *Crying: The Mystery of Tears*. New York: Winston Press, 1985.

Fukuyama, Francis. *Trust*. New York: Free Press, 1996.

Fulmer, Robert M., and Marshall Goldsmith. *The Leadership Investment*. New York: Amacom, 2000.

"Gene Therapy." *The Wall Street Journal*, November 11, 2002, p. D3.

Goldsmith, Marshall, et al. *Global Leadership: The Next Generation*. Upper Saddle River, NJ: Financial Times Prentice Hall, 2003.

Goleman, Daniel. *Emotional Intelligence*. New York: Bantam Books, 1995.

Green, Elmer. *Beyond Biofeedback*. New York: Delacourt, 1977.

Halberg, Franz, www.msi.umn.edu/~Halberg/

Helmuth, Laura. "Stem Cells Hear Call of Injured Tissue." *Science*, November 24, 2000, p. 1479.

Houston, Jean. *Jump Time: Shaping Your Future in a World of Radical Change*. New York: Jeremy P. Tarcher, 2000.

"Hot on the Trail of the Earwax Gene." *Science Now*, June 7, 2002, bric.postech.ac.kr/science/97now/02_6now/020607b.html

"Is It Your Smile? Your Laugh? Or Your Armpits? The Frustrating Science of Finding Pheromones." *Discover*, July 2003, p. 57.

Johnson, Spencer. *Who Moved My Cheese?* New York: G. P. Putnam, 1998.

Jung, Carl. *Memories, Dreams, Reflections*. New York: Vintage Books, 1989.

Kaye, Beverly, and Sharon Jordan-Evans. *Love 'em or Lose 'em*. San Francisco: Berrett-Koehler, 2002.

Leadership, www.pfdf.org

Maslow, Abraham. *Motivation and Personality*. New York: Harper-Collins, 1987.

Nash, Michael R. "The Truth and Hype of Hypnosis." *Scientific American*, July 2001, pp. 47–55.

Organizational Management, www.pfdf.org

Pinchot, Gifford, and Elizabeth Pinchot. *The End of Bureaucracy & the Rise of the Intelligent Organization*. San Francisco: Berrett-Keohler, 1993.

Pinchot, Gifford, and Ron Pellman. *Intrapreneuring in Action*. San Francisco: Berrett-Koehler, 1999.

Randerson, James. "Stem Cells Fix the Damage." *New Scientist*, January 11, 2003, p. 14.

Rogers, Carl. *On Becoming a Person*. New York: Mariner Books, 1995.

Seligman, Martin. *Learned Optimism*. New York: Alfred A. Knopf, 1991, www.positivepsychology.org/

Singh, Pradeep K., et al. "A Component of Innate Immunity Prevents Bacterial Biofilm Development." *Nature*, May 30, 2002, pp. 552–555.

Skinner, B.F. *About Behaviorism*. New York: Vintage Books, 1976.

"Staying Alive with Attitude." *Science News*, July 27, 2002, p. 53.

Stoltz, Paul G. *Adversity Quotient: Turning Problems into Opportunities*. Hoboken, NJ: John Wiley & Sons, 1997.

Surowiecki, James. *The Wisdom of Crowds*. New York: Doubleday, 2004.

"The T-shirt Test Tells." *Science*, April 20, 2001, p. 431.

Teamwork, http://reviewing.co.uk/toolkit/teams-and-teamwork.htm

Tears, www.cyquest.com/motherhome/healing_power_of_tears.html

Thomson, James, www.news.wisc.edu/packages/stemcells/

Thornhill, Randy, http://biology.unm.edu/Biology/Thornhill/rthorn.htm

Walsh, Larry. "Mirror, Mirror on the Wall, Who Is the Most Symmetrical of All?" *Quantum*, www.unm.edu/quantum_spring_2000/mirror.html

Wright, Karen. "Times of Our Lives." *Scientific American*, September 2002.

Yogi, Maharishi Mahesh. *Science of Being and Art of Living: Transcendental Meditation*. New York: Meridian, 1995.

Yunus. "The Grameen Bank." *Scientific American*, November 1999, p. 114.

Conclusion

601 The Universal Technologies

Coontz, Robert, and Phil Szuromi. "Issues in Nanotechnology: Taking the Initiative." *Science*, November 24, 2000, pp. 1523–1544.

Drexler, Eric. *Engines of Creation*. New York: Anchor Press, 1981.

Glenn, Jerome C., and Theodore J. Gordon. *2004 State of the Future*. Washington, DC: American Council for the United Nations University, 2004, www.stateofthefuture.org

"Holograms in Motion." *Technology Review*, November 2002, pp. 48–55.

"Hydrogen Fuel." *New Scientist*, October 5, 2002, p. 14.

Institute for Alternative Futures, www.altfutures.com/index.asp

Johnson, Steven. "Emerging Technology: Imagine if SimCity Wasn't Just a Game." *Discover*, May 2003, pp. 28–29.

Langreth, Robert. "Science beyond Belief." *Forbes*, December 23, 2002, pp. 316–318.

Liu, C., et al. "Hydrogen Storage in Single-Walled Carbon Nanotubes at Room Temperature." *Science*, November 5, 1999, pp. 1127–1129.

Martin, Richard. "The God Particle and the Grid." *Wired*, April 2004, pp. 114–117.

"Nanotubes." *Scientific American*, December 2000, pp. 62–66.

Zolli, Andrew, ed. *TechTV's Catalog of Tomorrow*. Indianapolis: Que Publishing, 2002.

602 The Best Answer—A Hybrid TechnEcology?

NASA Near Earth Object Program, http://neo.jpl.nasa.gov/
Sky City, www.takenaka.co.jp/takenaka_e/superhigh/2skycity/skycity.htm

603 "Tyger, Tyger, burning bright . . . "

Barker, Joel A. "Preparing for Change in the 21st Century," presentation
quoted in Dave Stinson, "Changing the Nature of Business,"
http://ppo.intergraph.cm/insight/1-1/1-business.asp
Blake, William. *Songs of Innocence and of Experience*. Princeton, NJ:
Princeton University Press, 1991.
Boorstin, Daniel. "The Fertile Verge: Creativity in the United States: An Address Given at the Carnegie Symposium on Creativity, November 19–20,
1980," www.dynamist.com/tfaie/bibliographyArticles/boorstin.html
Erickson, Scott W. "The Transition between Eras: The Long Wave Cycle." *Futurist*, July/August 1985, pp. 40–44.

Five Regions of the Future Publications List
(in order of importance)

- *The New Scientist*
- *Popular Science*
- *Mother Earth News*
- *Science News*
- *The Economist*
- *World Watch*
- *The Christian Science Monitor*

- *Nature.com*
- *Atlantic Monthly*
- *Tufts Health Letter*
- *Business Week*
- *Popular Mechanics*
- *Technology Review*
- *Fast Company*
- *Popular Mechanics*
- *Scientific American*
- *Mother Jones*
- *Industry Week*

- *Wired*
- *The Wall Street Journal*
- *Business 2.0*
- *Fortune*
- *Science*
- *The Futurist*
- *Utne Reader*
- *Newsweek*
- *Yes!*
- *Forbes*
- *New York Times Sunday Edition*
- *Invention & Technology*
- *Foreign Affairs*
- *Discover*
- *Air & Space*
- *Motor Trend*
- *Car and Driver*
- *EAA*

Index